A PRIMER OF
ROTATIONAL PHYSICS

3219

A PRIMER OF ROTATIONAL PHYSICS

Myrna M. Milani/Brian R. Smith

FAINSHAW PRESS

WESTMORELAND, NEW HAMPSHIRE

FIRST PRINTING

This book is manufactured in the United States of America. It is designed by James Brisson of Williamsville, Vermont and published by Fainshaw Press, Olde Spofford Road, Westmoreland, New Hampshire 03467.

Library of Congress Cataloging in Publication Data

Milani, Myrna M.
A primer of rotational physics

1. Physics. 2. Unified Field Theories. 3. Science—Philosophy. I. Smith, Brian R., 1939– II. Title.
III. Title: Rotational physics.
QC21.2.M56   1984   530.1   84-13518
ISBN 0-943290-01-5 (PBK)
0-943290-02-3
0 9 8 7 6 5 4 3 2 1

DEDICATION

To our Norwich University physics
class,
especially Tyrone Johnson
who kept us honest.

Contents

1 | Introduction to The Primer

Have you ever wondered why there is no unified explanation of the human race and its universe? Many reasons for this might pop into your mind:

1. There is no such thing.
2. There is a unified theory but it will take us thousands of years to find it.
3. There is a theory, but only scientists can understand it.
4. There are certain pieces of knowledge that humans shouldn't know.

Let's begin with the basic assumption that there is such a unified theory and that it's available to all who want to know it. However, if we look around, it's quite obvious no idea of unity is currently present or accepted. Even if we examine one field of human endeavor we find significant disunity within that field, not to mention the fact that the field has little relation to other fields.

For example, although great technical strides have been made in certain segments of Western medicine—the vaccines for diptheria and polio, for example—in many areas we have more superstitions about treating and curing physical ailments than the most primitive tribes. Does radiation, chemotherapy, or electric shock seem like a natural way to treat people who are sick? What about unnaturally keeping a patient alive? We're only beginning to recognize that diseases don't just happen to people. This is a difficult concept for many to accept because the obvious conclusion is that disease is a choice on the part of each individual. And what about all the emphasis on lengthening lifespan—has anyone questioned, really questioned, the reason for living longer? Is it in and of itself a "good" thing? Or does it indicate a wide spread belief in technological societies that *this* is all there is?

Has all this technology helped produce a healthier race? Has it improved our physicians' ability to diagnose diseases? Some recent evidence suggests doctors make the same percentage of misdiagnoses they made fifty years ago.

Because no explanations have been forthcoming regarding nontechnological healing, many Western physicians still have difficulty with faith healing, remission, or even acupuncture. Like many fields and professions, they consider these "in-house" problems or disunities, and rarely establish strong inter-relationships with disciplines such as physics or philosophy to help resolve them.

The medical field is in no way unique. Lawyers, accountants, cosmetologists, waitresses and longshoremen are as likely to create their own private world having their own special problems which are perceived as quite separate from all other areas.

So the question becomes—"why?" A story is told in academic circles about the college freshman who opens his final exam in Philosophy 101 and discovers there's only one question on the exam—"Why?" He gave the only answer he could think of—"Because." We could come up with lots of explanations why medicine is the way it is—or religion, or science, or psychology, accounting or cosmetology for that matter. The reality of the matter is they are what they are—but they don't have to stay that way.

We all want answers. Suppose we tell you it's possible to supply all our energy needs without depleting *any* resources such as oil or coal, without creating unwanted waste like ash or radioactive material, without having to deal with potentially dangerous forms of energy production, all cheaply and reliably. Does this sound outlandish or does it strike a responsive chord within you that sounds natural? Much of the answer depends upon your beliefs.

Even if the concept of universally available energy may seem farfetched for you, wouldn't you like some answers about yourself—where you came from, why you're here, where you're going? Although some people fear the answers for their own reasons, the answers are nonetheless available.

Regardless whether our questions are philosophical or scientific in nature, what most of us seek is some sort of unifying thread, something that links catalytic converters, food processors, nuclear reactors, sunsets, music, poetry, beagle puppies, mushrooms, gods and humans. Is it possible to create a Unified Field Theory (or better, Unified Fields Theory since, by it's very nature, it embraces *all* fields of human endeavor) as Einstein postulated? We think it is, if for no other reason than we can't think of any reason why it wouldn't be. So, the next question is: Where do we begin? The nice thing about any Unified Fields Theory (UFT) is we can begin anywhere, just as you can begin mowing your lawn anywhere, because by definition a UFT must include everything. When everything is present and all fields are related in one way or another, our first field of study becomes a matter of choice.

We chose to begin in the field of physics for a number of reasons:

1. A new, understandable theory of physics can lead the way to an entirely new era of development in machines and processes that are helpful to the human race, as well as lead to new philosophical awareness.
2. The time is right for a change in physics. If we follow the current path of science and technology, about all we're going to have is more of the same, much of it incomprehensible to the average person.
3. Once scientists and non-scientists alike grasp the concepts of rotational physics, we can apply these principles to all areas of our lives.
4. Physics is fun and challenging.

Using thought experiments (which you can do as you ride to

and from work, wait for kids at the orthodontist, or during inter-missions at concerts or sporting events) we present a theory of a smallest entity or unit from which everything arises, including thoughts and emotions. If the theory is logical, then what holds for that entity will hold for all that is. Although many of us recognize that the model of an atom with its electrons whirling about the nucleus looks like planets orbiting a star, so far there's been no integrated theory relating the two. Therefore, our UFT must in-terpret a Biblical phrase scientifically, "For inasmuch as you do it unto the least, you do it unto all." Whatever is true for the tiniest sub-atomic entity must be true for the largest spiral nebula and everything between.

In order to fulfill the criteria, we propose that all *(ALL)* is rotational—time, the human species, galaxies, rocks, molecules, atoms—everything. All is built from one basic rotational entity we call the *sub-seven alpha* and designate in shorthand like this:

$$_7\alpha$$

In the chapters ahead, we'll set the stage with some introduc-tory material, then follow the $_7\alpha$ from outer space through our atmosphere to the surface of the earth and see how it changes in response to the world around it. In this way we relate the concepts to you and what you already know about yourself and your world. Few people have the desire or patience to sit through abstract discussions of an invisible entity in a theoretical environment, so as we go along we'll make parallels of $_7\alpha$ activity in terms of things we know—billiard balls, skaters, rocks and strings, politicians, pro-nuke and anti-nuke advocates. In this way the theory will make sense right here and right now.

If you're like one of us, you may find the sight of an equation, Greek letter or exponent makes you shudder. Don't let it bother you. In each chapter the mathematics is minimal, but also followed by a verbal analogy. If the scientific material seems alien, don't get angry or frustrated; merely skim it and move on to the examples.

If the symbols bother you, mentally refer to them as squiggles, as in "The sub-seven squiggle," or "One squiggle is the same as the relationship between two doves and a hawk." After all, how many times do the more scientifically-oriented make their way through Dostoyevsky's classic novels by either skimming over or substituting other names for the main characters' tongue-twisting monikers?

The importance of rotational physics lies in its explanations or relationships among all things, not absolute values. If it's easier for you to relate to the book using mathematics, fine; if you prefer squiggles, skaters, strings and rocks, those relationships are every bit as valid and workable.

We each create our own reality. You may have heard this statement before just as it is, or in one of its many other forms:

Seeing is believing.
You are the light of the world.
You get what you concentrate on.
Everything is relative.
One man's floor is another man's ceiling.
You are the center of your universe.

We stress this point because regardless what we discuss, each individual's perception is critical. For those seeking absolutes or someone to tell them *specifically* how to view someone or something, they won't find that information here. Each person's reality is unique because each person is unique. You don't see things the way anyone else does; that *should* be an empowering concept rather than a limiting one. Because we're composed of the smallest entity, each 7α must also create its own reality and, like us,

1. Be conscious.
2. Have free choice.

Do these two qualities seem beyond comprehension? They shouldn't. Suppose everything in the universe possesses consciousness; the idea should be enlightening rather than restrictive. After all, you're conscious so why shouldn't the smallest unit embodied within you share that same quality? How can a conscious being of free choice be composed of anything less?

Now let's take a look at how perceptions affect all things, especially physics.

2 | Atoms and Other Small Things

We've all heard of atoms. By classic definition they're the building blocks of all matter or mass. Imagine a brick wall:

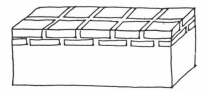

The wall is composed of bricks with mortar holding the bricks together so the wall doesn't lose its "wallness" and become just a pile of bricks; the bricks are the building blocks of the wall. If you consider a brick the final unit (i.e., nothing is smaller), then our brick wall describes the concept of the "building up" of matter. In our illustration we ignore the fact that a brick may be broken up into smaller pieces, because then it's no longer a brick but rather pieces of reddish, rock-like material. If you had a half-brick you might be able to contemplate what a whole brick looks like, but you wouldn't have the form of a brick. Only in the whole brick do you have full brickness that can be used to build a wall, a house, or any other structure. Among its many potential uses are

1. You could put one in your toilet tank to save water.
2. You could heat several and put them at the foot of your bed on a cold winter's night.
3. One brick could serve as a counterweight to help open a door.
4. You could throw it at someone or something.

The brick, then, has many uses beyond the usual ones connected with building a wall.

So it is with atoms. They can join together to form matter or they can do and be other things *as they choose*. These atoms do follow certain principles, but the principles more correctly describe what the atoms do; in other words, the principles *follow* the atoms. For example, suppose out of the hundred employees of the Mammouth Computer Corporation, eighty live in the same apartment complex. Although each one of those eighty chooses to live there for his or her own reasons, we could say, "Eighty percent of the Mammouth Computer Corporation employees live in Mammouth Condo Heaven." If you, as a new Mammouth employee, don't understand *where* this "principle" originated, you might believe you *had* to live in Mammouth Heaven. This isn't to say scientific principles don't have value. It is to say we must know where they came from, and that they in no way violate the free will and consciousness of each individual. Even if 100% of the Mammouth employees choose to live in Mammouth Heaven, that in no way decreases any individual's freedom to choose.

The word "atom" has been around a long time. It comes from the Greek *atomos* which means indivisible. The ancient Greeks understood the basic concept of matter built of very small indivisible pieces. Shakespeare may have been an early visionary of nuclear power when he spoke in *A Midsummer Night's Dream* about a small chariot driven by a "team of atomies." When the whole nuclear business got going earlier in this century, physicists believed they had found *the* atom and we're all familiar with the picture:

Scientists believed the atom had a nucleus or center made up of protons and neutrons with electrons whirling around it. This was a pretty good concept until they decided there might be more to the atomic structure than just protons, neutrons and electrons. Because their collective majority beliefs supported this idea, they constructed experiments to prove this and so discovered a whole bagful of particles which some wags have dubbed "winos, albinos, bozos, and sleptons." Now they're searching for this thing called a quark* which is supposed to be the ultimate, the smallest particle. The question is: If they find this elusive quark, what happens to the other 200 or so pieces of nuclear/atomic stuff they claim are around? If the quark's the smallest and they couldn't "see" it, isn't it just possible that some of those larger particles are made up of these smaller ones? It's gotten so bad that most scientists engaged in this business have to keep a directory nearby: can't tell the players without a scorecard.

There's a lesson in all this and it's this—we get what we concentrate on. As physicists concentrate on the belief there are more and more particles and semi-particles, they construct experiments designed to prove this belief. In short, they seek to prove themselves right and succeed. Actually what they find are variations on a theme, not a bunch of totally different entities.

For example, suppose we construct a maze of tinkertoys and decide to create experiments and equipment to break it apart. If we're good, we can design a machine accurate enough to blast the same clump of tinkertoys off the total mass most of the time. We can then use our experiment to *prove* the mass is composed of at least one part which looks like that clump. Someone else can come along and design equipment to consistently blast off another clump which may be larger or smaller than ours and even include some of the tinkertoys in our clump. However, because the clumps have different locations and configurations, we could say they're completely different entities. Now our original mass which is con-

---

*The term comes from Joyce's *Finnegan's Wake*—"three quarks for Mr. Marks."

structed of identical units, suddenly has two distinct "particles."
Depending on the location of the machinery and its target area, we
could consistently create and prove the existence of countless particles in our homogenous mass.

Part of the problem lies with faulty or incomplete assumptions,
and one of these has to do with vibration or pulsation. Because this
is a book on rotational physics, let's start with the difference between
rotation and vibration.

Imagine a rock tied to a string that you're whirling around your
head. To a bird in a tree looking down, you look like this:

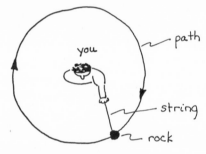

It's perfectly obvious to the bird what you are doing. Let's add
some measurement to what's going on and say the rock makes a
complete circuit every second. Therefore, you're producing a rotational rate of one revolution per second (rps) or sixty revolutions
per minute (60 rpm).

Now suppose I come walking towards you with my eyes at the
same level as your rock-string system. When I watch what you're
doing, I don't see a circular path at all:

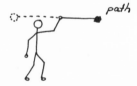

What I see is a back-and-forth motion which *oscillates* at a rate of

60 oscillations or cycles per minute.* The bird in the tree sees rotation whereas I see oscillation or vibration. Although what I see is correct for me from my point of observation, I'm not seeing the bigger picture. I could create an entire body of thought centered around the "fact" that when someone puts a rock on a string and then applies a certain motion to the string with their arm, it produces vibration.

Now let's confound our system a little bit further by introducing some magic. Magic occurs in physics when we see the final outcome but not the process which led to it. Let's assume our rock-string system can disappear and re-appear at will. For very short disappearances (less than a tenth of a second every revolution, for example), we could still determine the path of the rock. From our bird's point of view it would look like this:

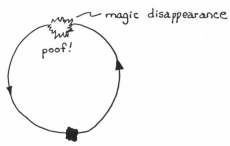

However, suppose the rock disappears for an entire hour and then re-appears for only a tenth of a second before it disappears again. Wouldn't it be easy to be entirely misled about what was going on? If we were to hang around for an entire workday (8 hours) we might decide we saw eight *different* rocks and not the same one at all.

Because small particles or entities rotate rather than vibrate and because they seem to disappear right in front of those who are watching them, we often identify the same "rock" as a massive

---

*This could also be expressed as 1 Hertz. A Hertz (Hz) is not only a car rental company, it also represents one cycle (oscillation) per second.

number of different kinds of tiny "rocks." However, there is only one basic entity from which *everything* comes—heat, light, sound, electrical potential, mass, pressure—the $_7\alpha$ (sub-seven alpha). Let's begin with a discussion of the physical properties of this phenomenon.

3 | The Sub-Seven Alpha $(_7\alpha)$

Notice we don't speak of the $_7\alpha$ as a particle; we speak of it as an entity. There are two reasons for this. In the first place, the use of the word "particle" usually calls to mind a physical piece of matter like a particle of dust. Physicists still speak of very small sub-atomic elements as particles even though they know they're not talking about miniature marbles. However, the use of the term really becomes confusing when it's used to describe certain phenomena such as light, which seem to behave sometimes as a wave,

and sometimes as a particle.

Zoom!

But as we'll see in chapter nine, light is neither; it's composed of a unique combination of $_7\alpha$.

The second reason we use the word entity is because "particle" seems to denote a rather dead, impersonal hunk of something. The $_7\alpha$ is a single unit possessing its own form of consciousness. We could have just as easily chosen the word "being" but we prefer to use "entity."

If we ask you to describe yourself in as great detail as possible, we would get many simultaneous definitions. You have a physical size (height, weight), an occupation, an age, a family status, a sex, a position in the community. For example, it's not difficult for you to see yourself as a 35-year old, slightly overweight, accountant, father. So it is with the $_7\alpha$; it can also be several things at once. In one sense it can be mass and energy *at the same time,* but appears to function primarily as one or the other. It always has the potential to be either. When it functions in the mass role, it still retains some "energyness" just as our accountant, while at work, still possesses some "fatherness."

To introduce the concepts of rotational physics, we need to start with a diagram of a spiral:

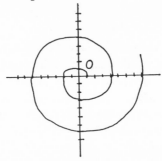

Note that in each quarter revolution, the path of the spiral moves one more unit away from point O which is called the origin. This particular spiral is called the Spiral of Archimedes and it's a common shape in nature. The chambered nautilus sea shell is one example, spiral galaxies are another. Imagine an object moving along the path of the spiral. We can see this easily if we imagine a man swinging a rock around his head like you did in chapter two, creating a rotational speed of one revolution per second. Now let's complicate life for our friend who may be practicing to use a slingshot similar to the one David used to fell Goliath. First we tell him to make little knots in the string every six inches:

Then, after he begins swinging the rock around his head, we tell him to play out one knotted length of string (6 inches) every quarter turn or 90 degrees. After the first complete revolution, the rock is four units (two feet) away from him; at two revolutions, the rock is eight units or four feet away. The rock moves another two feet away every second and we can express that in the formula

$$d = 2t$$

where d (distance) is in feet when t (time) is expressed in seconds. If our person could physically perform this feat, after forty-four minutes, the rock would be one mile away!

There's also another effect going on here. Because the rotational rate of one revolution per second remains constant, the linear speed of the rock is constantly increasing. Because the distance traveled by our man's arm is much less than that the rock at the end of the string must cover *in the same amount of time,* the rock must go much faster. This is merely a variation of crack-the-whip; the person in the center hardly moves at all whereas the one on the end has to go like crazy to keep up. At a distance of one mile out, the rock would be traveling at 22,600 mph or 6.3 miles per second—about 12 times the speed of a rifle bullet! Now we can't see a rifle bullet and our person—or anyone else for that matter—certainly couldn't see the rock. What he would see from his vantage point is a string going away from him which at some point disappears.

Because the rock can't be seen, we could say it's no longer "there"—it's disappeared. We could also say the rock is no longer totally a visible mass form at all but has now assumed a more invisible energy form. The rock still retains some "massness" but it's become much more of an energy form than it was at rest. So in our thought experiment we can say that the further out the rock is from the person holding the string, the more the system operates as energy rather than as mass. If we need an arbitrary point to determine when the rock becomes more energy than mass, we could say that this occurs when the rock can no longer be seen. This may sound quite strange, especially when we consider each person has a slightly different perception of when the rock disappears. That means each person must recognize a different mass-to-energy conversion point; a near-sighted person sees it disappear sooner than someone with 20-20 vision. Well, that's one of the ways rotational physics differs from traditional physics. It's human-dependent, not absolute. We each create our own reality and that includes the phenomena that occur in the field of physics.

Now let's imagine our person grows quite tired whirling this rock around his head and wants to call it quits. He lets the rock come back simply by winding the string around his hand. Soon the rock reappears and begins to slow down. Finally, our experimenter winds up with a ball of string and a rock which no longer moves. When the rock comes to rest, we say we have reached the point of *maximum mass*. That's obvious because our person has all this physical mass at one point.

Let's look at another concept. Our rock-slinger replaces the rock with an eight ball which has a small shaft placed in a hole drilled through the center of the ball.

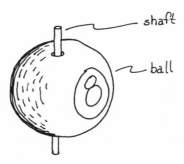

Next, he attaches string to the ends of the shaft and gives the ball a very rapid spin while holding it in his hand.

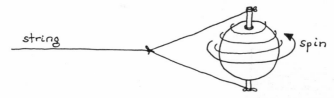

Our experimenter once again plays out the string and rotates the system as before. Now the ball moves like a planet revolving around a sun—it has both rotation and spin. As the ball begins to move further and further from the point of origin it gains linear velocity (still using a rotational rate of one revolution per second) but the spin rate decreases. In our "real world" this spin reduction is due to friction.

If we paint a bright dot on the ball and if our bird in the tree has not become bored with all this human activity and flown off, our avian friend sees a pattern like this:

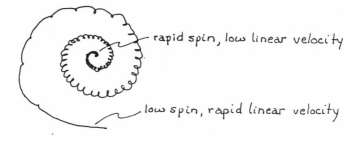

When the string is played out quite a distance, the linear velocity is very high but the spin rate is much slower than when the ball is held by the experimenter. We can say that some spin is "given up" to velocity; that is, spin slows as linear velocity increases.

To bring the ball back we use a magic system which decreases the linear velocity as the spin starts increasing again. Think of a figure skater about to do a series of in-place pirouettes. The first thing the skater does is build up some linear (straight-line) speed. When she *feels* (please note the use of the word here), when she *senses* the time is right, she converts all that linear speed to spin.

Let's go back to the ball. As the ball comes in closer and closer to the origin, its spin increases until, when it sits at rest once more, it's spinning at its original rate.

This describes the most basic activity of the $_7\alpha$ entity. These entities traverse a rotational path that is a Spiral of Archimedes, but there are some differences between the $_7\alpha$ activity and our rock-string system. In the first place, there's no string. Secondly, the $_7\alpha$ is not a hunk of "something." The entity itself has no mass and doesn't "become" mass, but behaves in a manner that *confers* massness when it's spinning more than translating.* As it moves outward along the spiral, the spin rate decreases. The third major difference between the rock-string and the $_7\alpha$ concerns the nature of the spiral. The $_7\alpha$ has no origin or beginning point within the limits of our present discussion. If we have a spiral,

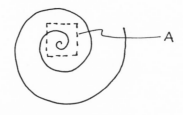

---

*Translation is an engineering or scientific term that refers to the linear movement of an object without rotation. Every point on a translating body keeps the same spatial relationship with all other points.

and enlarge portion A, what we see as an origin or end point,

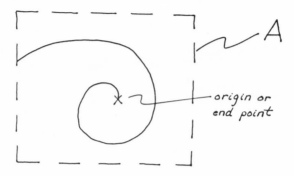

is really another spiral:

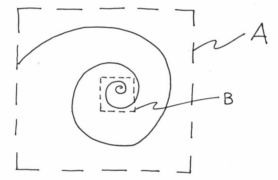

If we then enlarge portion B on the above diagram we see still another spiral, and another and another, ad infinitum.

This spiral path of the $_7\alpha$ helps explain simple things like the water funnels formed when you pull the plug on a full sink of water, or complicated things like the tracks of subatomic particles in bubble chambers or the nature of spiral nebulae in the universe.

Now let's relate the $_7\alpha$ and its properties to mass and energy. We don't want to belabor the point, but it's important not to think of the $_7\alpha$ as a tiny piece of some *thing*, meaning something like a small particle. Instead, think of it as a "probability mist." A probability mist diagram of the spiral track of the $_7\alpha$ (a small part of the track, anyway) looks like this

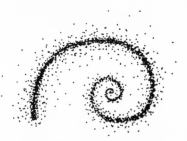

with the darker portion of the track representing places where the $_7\alpha$ is *more likely* to be found and the lighter, dotted area as places it's *less likely* to be. This isn't a new concept at all. Physicists, beginning with Heisenberg and his Uncertainty Principle—which states that you can't be totally definitive about both the position (mass) and momentum (energy) of an electron; the more you know about one, the less you know about the other—have long spoken of the probability mist that seems to surround the classic particles such as protons, neutrons and electrons. The $_7\alpha$ does have size to it, but even that is changeable depending upon the state of the $_7\alpha$. In fact, it's easier to think of the $_7\alpha$ having function rather than size or shape; it may function as energy, be perceived as mass, or create sound, light, heat or pressure. The average diameter of the $_7\alpha$ *field of influence* is 0.0000000000001 millimeter. We say average because it can vary substantially.

Let's pause here and talk about very small and very large numbers because we'll be dealing with them from time to time in our discussion. We said the average field or sphere of influence of the $_7\alpha$ is .0000000000001 mm. That's quite a small number. In fact, it's about 100 times smaller than the diameter of an average atomic nucleus—that part of the atom that contains protons and neutrons but excludes the orbiting electrons. To give you another idea of how small the $_7\alpha$ is, if we lined up $_7\alpha$'s like this

we'd have 254 trillion of them in one inch. If we made each $_7\alpha$ one inch in diameter, 254 trillion of them would cover the distance from the earth to the sun 43 times!

From now on we'll use what is called scientific or exponential notation which is just a shorthand that helps us write big or small numbers. Here's how it works.

The number 100 can also be expressed as 10 times 10 or 10 squared. The shorthand for 100 is written 10^2. The 2 above the 10 is called an exponent and designates the number of tens to be multiplied. In like manner, 10^3 is $(10 \times 10 \times 10)$ or 1000. The table below gives the scientific notation for some common numbers:

Name	Written Form	Scientific Notation
Ten	10	10^1
One hundred	100	10^2
One thousand	1,000	10^3
Ten thousand	10,000	10^4
One hundred thousand	100,000	10^5
One million	1,000,000	10^6
One billion	1,000,000,000	10^9
One trillion	1,000,000,000,000	10^{12}

In scientific notation, the number 254 trillion is written

$$254 \times 10^{12}$$

or

$$2.54 \times 10^{14}$$

which is the same thing.

Small numbers (those less than 1.0) work in the reverse. But what about the number 1 itself? If you look in our table, you notice that as the numbers get smaller, the exponents do likewise. But what is 10^0? Well, 10^0 is 1. As a matter of fact, any number to the "zero power" is 1. We aren't going to develop the reasoning behind this, but any math book discussing logarithms will explain it if you're curious.

It's no surprise that we use *negative* exponents for numbers less than 1.

Name	Written Form	Scientific Notation
One tenth	.1	10^{-1}
One one-hundredth	.01	10^{-2}
One one-thousandth	.001	10^{-3}
One one-ten-thousandth	.0001	10^{-4}
One one-hundred thousandth	.00001	10^{-5}
One one-millionth	.000001	10^{-6}
One one-billionth	.000000001	10^{-9}
One one-trillionth	.000000000001	10^{-12}

This handy rule of thumb may be helpful: if the number has a positive exponent (10^3), the exponent designates the number of zeros after the 1 (1000); if the exponent is negative (10^{-3}), that's the total number of "places" after the decimal point (.001).

We'll also use metric units in our discussions because metric measurement is the accepted scientific form:

1 meter (m) = 100 centimeters (cm) = 1000 millimeters (mm)

1 kilometer (km) = 1000 meters (m)

A meter is a little longer than a yard (actually, 39.37 inches); there are about 2.5 centimeters to an inch and a mile contains 1609 meters. The speed of light in the English system—186,000 miles per second—is 300,000 kilometers per second (km/sec) in the metric system, or written another way, 3×10^8 *meters* per second (m/sec).

Now that we understand scientific notation and metric units, let's go back to our spiral and the concept of spin versus linear motion.

A

As we said, when the ₇α chooses to exist in the realm of A, it experiences high linear motion and low spin. Does a ₇α ever get to the point where it's *all* linear motion and no spin at all? Although theoretically possible, we can argue that even the most linear path we can imagine—a straight line—

actually never occurs in our reality. For our purposes, though, let's assume it does. So when we speak of the ₇α in a state of 100% linear motion, we can say it is *all* straight-line motion with no spin. In this all-linear-motion state it's most energy-like. However, remember that the ₇α, like you, always retains the *potential* to be anything. It can be total linear motion,

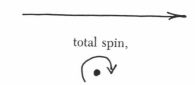

total spin,

or something between the two extremes.

This is akin to the artist who says, "Give me a line

and a dot

●

and I can create anything." So the $_7\alpha$, especially in combination with other $_7\alpha$ can create anything. The $_7\alpha$ can be everything and nothing; it can be everything in combination and nothing by itself.

It's the same with people. By ourselves, we're nothing, which doesn't mean that singularly we have no reality. Imagine yourself somewhere where there is nothing else: no-thing else. Do you have any meaning? Does it make any difference whether you write the most beautiful poem or solve the most complex equation? Does it matter whether you shave or set your hair? Whether you live or die? No: we need another to give ourselves and what we do meaning, relevance, importance. So it is with the $_7\alpha$; by itself it's only itself. This isn't just valid philosophical musing but also a very important physical concept.

If we can think of the $_7\alpha$ in its stretched-out form

$$\longrightarrow$$

as being in an energy state, what happens when it gives up linear motion in favor of spin? It slows down. As it slows down in its linear travel, it spins more. We can depict this graphically as

1.   2.   3.   4. 5.

As the $_7\alpha$ slows, it begins to spin or rotate more rapidly. In state 1, the $_7\alpha$ travels in a relatively straight line—it has high linear motion and practically no spin. In state 2, it has swapped some linear motion for spin. States 3 and 4 are simply more of this same transfer; when we reach state 5 the $_7\alpha$ is all spin and no linear motion.

What causes the $_7\alpha$ to change its form? Primarily, the choice of the $_7\alpha$ itself. Does this mean that there's no predictability, that it's a random change? No. When the $_7\alpha$ begins to encounter other $_7\alpha$, certain patterns of group behavior emerge. Imagine a crowded railroad station such as Grand Central Terminal in New York City during morning rush hour. Each person has free choice—to stay on the train, to buck the flow of commuters, to bump into other people—but there's a general predictability to the mass flow: an orderly filing out of the train, along the platform, through the terminal and out onto the streets.

We have another variation of the Heisenberg Uncertainty Principle—if we want to know everything about an individual entity, then we can know nothing about the group and vice versa. We've all experienced this phenomenon many times. If we concentrate on only a single plant, we have no idea what's happening in the rest of the garden. If we get caught up in the music and splendor of the All-State Band, we don't hear son Johnny's sour note.

Now that we know how an individual 7α functions, let's consider how it interacts with others by following a 7α as it comes from outer space to earth. To do this we must first postulate a series of concentric rings or "layers" that surround our earth.

4 | The Layers Above the Earth

As we proceed towards the surface of the earth, the concentration of $_7\alpha$ increases; and as the $_7\alpha$ become more closely packed, more reactions and interactions occur so that by the time we reach the surface of the earth, the $_7\alpha$ concentration (or density) is sufficient to create perceptual mass. In order to understand these changes, we're going to divide the area surrounding earth into layers and then examine what goes on in those layers and the areas between them.

Let's talk first about the layers themselves. Although not drawn to the proper scale, the following drawing shows 12 rings around the earth, each having a number which designates the *center* or mid-point of each layer. The center of each layer is located at a predictable distance from the surface of the earth, and each layer has a width or "thickness" such that its outer diameter is twice the distance from the earth as its inner diameter. Also the center of each successive layer going outward from earth (13, 12, 11. . .) is twice as far from the earth's surface as the previous one. In other words, if the center of layer 13 occurs at x distance from the earth's surface, then the center of layer 12 is 2x, layer 11 is 4x, and so on to layer 2 which is 2048x from the earth.

In this case, x is equal to 2500 meters or a little over a mile and a half above the surface. With that in mind, we can determine the distances from the surface of the earth to the center or mid-point of each layer:

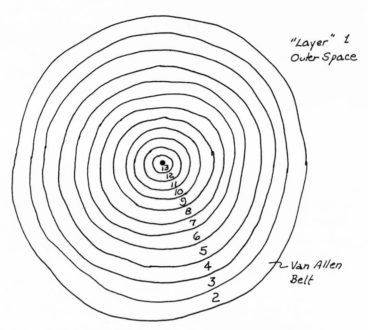

"Layer" 1
Outer Space

Van Allen
Belt

Layer No.	Distance in Meters
13	2,500
12	5,000
11	10,000
10	20,000
9	40,000
8	80,000
7	160,000
6	320,000
5	640,000
4	1,280,000
3	2,560,000
2	5,120,000 (a little over 3,000 miles)

These layers "orbit" the earth in concentric paths such that each layer has the same rotational rate or spin. This phenomenon is like a phonograph turntable; all points have the same *spin* (33-1/3 rpm for example) but two points at different distances from the center spindle have different *velocities*. To see this effect, place two coins on a turntable:

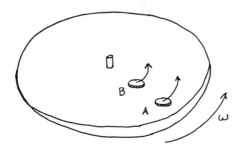

Even though the turntable turns (spins) at a constant rate (ω),* coin A travels linearly (translates) faster than coin B. If A is twice as far from the spindle as B, its linear velocity (speed, motion, translation) is twice as fast. Similarly, because the earth's layers have the same form of rotation, to an observer standing on the surface of the earth the layers appear fixed.

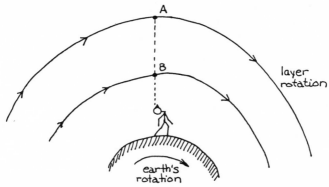

*The symbol ω is the Greek letter "omega" which is classically used in science and engineering to denote rotation or spin in rpm or rps.

In other words, if our observer could fix points A and B within any two of the layers, A and B would always stay in the same position relative to one another.

Each layer has its own width or thickness. We already said each layer except outer space (layer 1) is constructed such that its outer fringe is twice the distance from the earth as its inner fringe. Knowing this, can you figure out the width of each layer? You could do this either deductively or mathematically, but we'll do it mathematically. Pick a layer, any layer. We already know two things about it:

1. The distance of its center, or mid-point (M) from the surface of the earth (y);
2. The fact that its outer fringe (A) is twice as far from the surface as its inner border (B); hence the mid-point (M) is equidistant from either fringe.

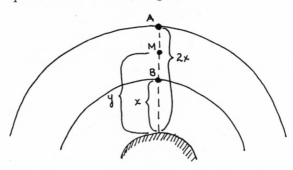

This means that AM equals MB. If B is x meters from the surface of the earth, A is 2x meters. Then we can say that

$$AM = MB$$
$$2x - y = y - x$$

for any layer. We can pick any layer we want because we know what y is for all layers. Let's pick layer 11 for which y equals 10,000 meters:

$$2x - 10,000 = 10,000 - x$$
$$3x = 20,000$$
$$x = 6,667 \text{ meters}$$
$$2x = 13,333 \text{ meters}$$

So layer 11 looks like this:

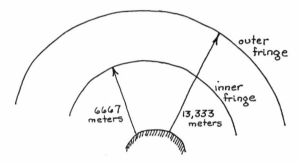

and you can see the two-fold distance difference. The mid-point is at 10,000 meters and each fringe is 3,333 meters from it.

Now let's do the same calculation for layer 10 whose mid-point is 20,000 meters from the earth's surface:

$$2x - 20,000 = 20,000 - x$$
$$3x = 40,000$$
$$x = 13,333 \text{ m.}$$
$$2x = 26,667 \text{ m.}$$

Here we see that the outer fringe of layer 11 coincides with the inner fringe of layer 10; they are both 13,333 meters from the earth's surface. The area where the outer fringe of layer 11 meets the inner fringe of layer 10 is known as an *interphase*. As we shall soon see, interphases have special qualities which differentiate them from layers.

Now let's discuss two characteristics of $_7\alpha$ concentration. As we proceed towards the earth, the concentration of $_7\alpha$ within each successive layer increases. This occurs because the *number* of $_7\alpha$

within each layer is the same but the amount of space to hold them decreases. Secondly, the concentration within any layer is greater at the inner fringe than at the outer fringe.

Imagine a circular race track 1,000 meters long (in circumference):

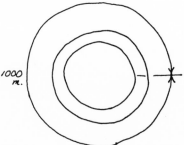

1000 m.

Now put 1,000 evenly spaced runners on the track with each runner 1 meter from the person in front and the person behind:*

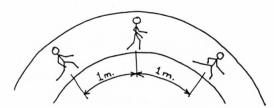

As long as our runners maintain their 1-meter spacing, there's plenty of room for any runner to stretch his or her arms out, jump into the air, possibly even stop for a split second (as long as they make up the lost distance). Now let's cut the circumference of the track to 500 meters. Each runner now only has a 1/2 meter space to the front and rear.

*Assuming, for a moment, that the runners have no "width."

Because they're closer together, the runners are "denser", more concentrated. Instead of an average of one runner per meter, we now have two runners per meter. If we keep reducing the circumference, it isn't long before the runners are so tightly packed they interfere with one another:

Anyone who drives during the rush hour knows what happens—increased concentration causes a slowdown. This is true for runners, for cars, and for 7α.

So, as we go from one layer to the next we double the density of 7α.

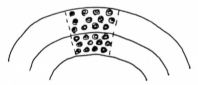

As the density (concentration) increases, total velocity (V_T) decreases. What's the magnitude of this decrease? In the upper layers $(2,3,4,5. . .)$ the decrease occurs by a factor of ten. Therefore, if the average total velocity in layer 2 is 10^{26} m/sec, in layer 3 it is 10^{25} m/sec.

Before we continue with our layer-to-layer discussion, let's stop and address what some of you already see as a gross inconsistency. We said the layers rotate concentrically (evenly) in relation to an observer on the surface of the earth. If layer 2 is *twice* as far from the earth as layer 3, then why a *ten-fold* velocity difference? Worse yet, the surface of the earth moves quite slowly (a little more than 400 m/sec as a result of its own spin). How can we ever reconcile such a vast difference between 4×10^2 m/sec and values as large as 10^{25} or 10^{26} m/sec?

The answer is that the $_7\alpha$ within a layer don't travel like this:

Remember, this is *rotational* physics and within the layers singular and groups of $_7\alpha$ travel a rotational path like this:

Their spiral motion then imparts a linear-like motion to the layer itself. Remember as a child playing in a cardboard box? You got inside and by crawling made the box move forward.

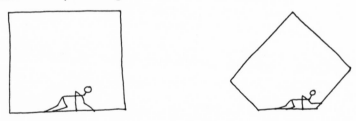

Or think of several children running inside a large piece of culvert pipe causing it to rotate and translate.

Now let's examine the interphase where one layer meets another. Remember: don't think of the layers as having distinct boundaries like that observable between a liter of oil and a liter of water in a container. Our layers aren't like that. The interphase is an area where one layer blends into another:

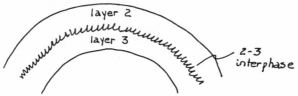

Let's come in from outer space again, through layer 2, and the 2–3 interphase and into layer 3. We can say two things about our incoming $_7\alpha$:

- It has an average total velocity (V_T) of 10^{26} m/sec.
- It's all linear velocity and no spin, therefore

$$V_T = V_S + V_L$$
$$10^{26} \text{ m/sec} = 0 + 10^{26} \text{ m/sec}$$

and the relationship between V_S and V_L expressed in per cent is:

$$\frac{V_S}{V_L} = \frac{0\%}{100\%} \text{ or just } \frac{0}{100}$$

If our $_7\alpha$ joins layer 2 as a typical inhabitant, its total velocity remains at 10^{26} m/sec but its V_S/V_L becomes 50/50 and:

$$V_S = 5 \times 10^{25} \text{ m/sec}$$
$$V_L = 5 \times 10^{25} \text{ m/sec*}$$

As the $_7\alpha$ exits layer 2 heading towards layer 3, it undergoes another change. Its total velocity remains the same, but it reverts to a V_S/V_L of 0/100 once again. Therefore, within the 2–3 interphase

$$V_T = 10^{26} \text{ m/sec}$$

---

*Half of 10^{26} is 5×10^{25}.

Coming into layer 3, then, our $_7\alpha$ is the same as it was in outer space. The most probable outcome, then, is that

- Its total velocity will decrease to 10^{25} m/sec.
- Its V_S/V_L will become 50/50 again.

Let's review the progressions:

Layer 1	10^{26} m/sec	0/100
Layer 2	10^{26} m/sec	50/50
Interphase 2–3	10^{26} m/sec	0/100
Layer 3	10^{25} m/sec	50/50

The reason it slows down is because of the greater concentrations of $_7\alpha$.

There's an important difference between layers and interphases regardless which ones we're talking about. The interphases don't travel in concentric circles like the layers. Their motion is perpendicular to the earth and stationary relative to earth's moving surface:

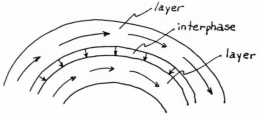

To visualize this, imagine you're at an airport having two separated moving walkways, one moving at 1 m/sec and another moving at 1/2 m/sec, and you're on the faster walkway:

If you want to change to the slower one, the fastest way to do this (i.e., make the change by covering the least distance) is to step off the 1 m/sec walkway and move perpendicularly towards the 1/2 m/sec walkway:

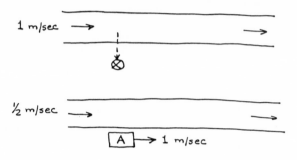

If we place an observer at point A in a little baggage cart moving at 1 m/sec, what does he see? Well, when you were on the 1 m/sec walkway, you had no motion relative to the observer. He saw you as stationary because both of you were moving at the same rate in the same direction. As soon as you step off, you suddenly appear to him to be moving *backwards* at 1 m/sec. Similarly, to an observer on the earth, the layers appear stationary whereas the interphases appear to move linearly when, in fact, just the opposite occurs.

Now let's return to our $_7\alpha$ progression towards the earth. When we last left our $_7\alpha$ it was a 50/50 moiety within layer 3. As it exits layer 3 and enters interphase 3–4, it retains its total velocity of 10^{25} m/sec but again reverts to an all linear velocity entity of 0/100. It then enters layer 4 and has an 80% probability of becoming a 50/50 once again, and loses another 1/10th of its total velocity. We can summarize these two changes in the same way as before.

| Interphase 3–4 | 10^{25} m/sec | 0/100 |
| Layer 4 | 10^{24} m/sec | 50/50 |

By now this may seem rather dull, boring, and routine. You're probably saying, "I know—when it enters interphase 4–5 it reverts to 0/100 once again at a total velocity of 10^{24} m/sec." Wrong. Actually, you're only half wrong. It's total velocity decreases to 10^{24} m/sec all right, but it doesn't revert to the 0/100 state. It becomes a 10/90! Again, remember that the concentration of 7α continues to increase as we approach earth and by the time we reach interphase 4–5 that concentration has become great enough that the 7α can't revert to the 0/100 form. There's no longer enough room for it to fully stretch out into its linear form, even with a reduction in total velocity.

In interphase 4–5, then, we have

$$V_T = 10^{24} \text{ m/sec}$$

and

$$V_S/V_L = 10/90$$

and therefore

$$V_S = 10^{23} \text{ m/sec}$$
$$V_L = 9 \times 10^{23} \text{ m/sec}$$

In the remaining interphases (5–6, 6–7, 7–8. . .) the V_S/V_L continues to change by 10% for both V_S and V_L until we near the surface of the earth where the ratio becomes all spin and no linear velocity or 100/0. In table form, the progression looks like this:

Interphase	V_S/V_L
5–6	20/80
6–7	30/70
7–8	40/60
8–9	50/50
9–10	60/40
10–11	70/30
11–12	80/20
12–13	90/10
Earth	100/0

Now let's drop rapidly from interphase 4–5 and layer 5 to layer 10 and see how the total velocities decline in our downward journey:

Layer	V_T (m/sec)
5	10^{23}
6	10^{22}
7	10^{21}
8	10^{20}
9	10^{19}
10	10^{18}

You should now be in a position to figure

- The distance from the earth.
- The spin velocity (V_S), for layers 2–11.
- The linear velocity (V_L), for layers 2–11.

of the

- Center of any layer.
- Edge of any layer—the interphase.

Try one and then check your answer using the table at the end of the chapter. If you don't get the right answer, go back and review the concepts because we're going to introduce a new twist as we come out of layer 10 and into interphase 10–11.

Because concentrations steadily increase, when our $_7\alpha$ leaves layer 10 and enters interphase 10–11, it loses total velocity *in the interphase*. This is the first time this phenomenon occurs. In the upper layers and interphases, total velocity remains the same when a $_7\alpha$ moves from a layer to an interphase and in the very upper layers it was able to revert to its outer space form of 0/100 when it got to the interphase.

When our $_7\alpha$ is within layer 10, 80% of the moieties are 50/50 configurations with

$$V_T = 10^{18} \text{ m/sec}$$

Therefore

$$V_S = 5 \times 10^{17} \text{ m/sec}$$
$$V_L = 5 \times 10^{17} \text{ m/sec}$$

Looking back at our table of configurations within interphases we see for interphase 10–11

$$V_S/V_L = 70/30$$

So we might expect that

$$V_S = 7 \times 10^{17} \text{ m/sec}$$
$$V_L = 3 \times 10^{17} \text{ m/sec}$$

but that's not the case. Within interphase 10–11,

$$V_T = 10^{17} \text{ m/sec}$$

and

$$V_S = 7 \times 10^{16} \text{ m/sec}$$
$$V_L = 3 \times 10^{16} \text{ m/sec}$$

From here to the surface of the earth, there's a drastic reduction in total velocity:

Layer or Interphase	V_T (m/sec)
11	10^{16}
11–12	10^{8}
12	10^{4}
12–13	10^{2}
13	10^{1}
Earth	10^{0}

So, that means the *surface* of the earth is an interphase with some rather special characteristics:

- Our V_S/V_L is 100/0—all spin and no linear velocity.

- V_T is 10^0 or 1 m/sec—about normal walking speed.
- The $_7\alpha$ concentration is so great we have something that can be seen and felt—mass.
- This interphase has an actual "thickness" to it.

Let's examine the matter of thickness. Earlier in the chapter we mentioned that the outer edge of one layer touches the inner layer of the one above it and it is at this point (Remember the "point" is really a sphere.) that the interphase occurs.

For example, the center of layer 11 is 10,000 meters from the earth. It's inner and outer fringes are 3333 meters on either side:

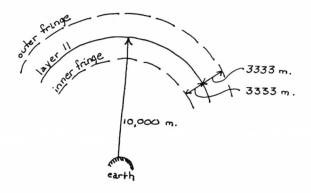

The outer fringe, 13,333 meters from earth's surface, is also the location of

1. Interphase 10-11
2. The inner fringe of layer 10.

We can now condense our math to a rule of thumb:

To find the distance from the surface of the earth to the inner fringe of any layer, double the height (altitude) of the center of the layer and divide by 3.

For example, to find the distance to the inner fringe of layer 13:

$$\frac{2500 \times 2}{3} = \frac{5000}{3} = 1667 \text{ meters}$$

So the interphase called EARTH extends above the surface 1667 meters. Because this is only half the width of the interphase, the other half extends 1667 meters below the surface. We're willing to wager that 80% of "life" (people, animals, insects, plants) exists within this two-mile band, and 10% exists above it, and 10% below. We'll even stretch a bit further and say that of the inhabitants in the interphase

- 10% are "passing through" (very young, very old).
- 10% are single.
- 80% are combinations of one form or another.

Think about this in terms of people. Does it seem to make sense?

Here's the summary table we promised you before we move on to more of the life and times of the 7α.

LAYER	V_T (m/sec)	DISTANCE FROM EARTH TO CENTER OF LAYER (meters)
1 (outer space)	10^{26}	—
2	10^{26}	5.12×10^6
3	10^{25}	2.56×10^6
4	10^{24}	1.28×10^6
5	10^{23}	6.4×10^5
6	10^{22}	3.2×10^5
7	10^{21}	1.6×10^5
8	10^{20}	8×10^4
9	10^{19}	4×10^4
10	10^{18}	2×10^4
11	10^{16}	10^4
12	10^4	5,000
13	10	2,500

$$\frac{V_S}{V_L} = \frac{50}{50} \text{ for 80\% of moieties except "layer" 1.}$$

INTER-PHASE	V_T (m/sec)	$\dfrac{V_S}{V_L}$ % Ratio	$\dfrac{V_S}{V_L}$ m/sec
1–2	10^{26}	$\dfrac{0}{100}$	$\dfrac{0}{10^{26}}$
2–3	10^{26}	$\dfrac{0}{100}$	$\dfrac{0}{10^{26}}$
3–4	10^{25}	$\dfrac{0}{100}$	$\dfrac{0}{10^{25}}$
4–5	10^{24}	$\dfrac{10}{90}$	$\dfrac{10^{23}}{9 \times 10^{23}}$
5–6	10^{23}	$\dfrac{20}{80}$	$\dfrac{2 \times 10^{22}}{8 \times 10^{22}}$
6–7	10^{22}	$\dfrac{30}{70}$	$\dfrac{3 \times 10^{21}}{7 \times 10^{21}}$
7–8	10^{21}	$\dfrac{40}{60}$	$\dfrac{4 \times 10^{20}}{6 \times 10^{20}}$
8–9	10^{20}	$\dfrac{50}{50}$	$\dfrac{5 \times 10^{19}}{5 \times 10^{19}}$
9–10	10^{19}	$\dfrac{60}{40}$	$\dfrac{6 \times 10^{18}}{4 \times 10^{18}}$
10–11	10^{17}	$\dfrac{70}{30}$	$\dfrac{7 \times 10^{16}}{3 \times 10^{16}}$
11–12	10^{8}	$\dfrac{80}{20}$	$\dfrac{8 \times 10^{7}}{2 \times 10^{7}}$
12–13	100	$\dfrac{90}{10}$	$\dfrac{90}{10}$
Earth	1	$\dfrac{100}{0}$	$\dfrac{1}{0}$

5 | From Dust Ye Came: A Journey Towards Earth

Let's imagine for just a moment that the earth is the only thing we know; we can then say that earth is our reality. Of course, reality is relative. An astrophysicist's reality may include the entire universe which some scientists believe is 10 billion light-years (9.5 × 10²⁵ meters) in diameter. To an African pygmy, or child of the forest as he calls himself, reality may only be a hundred square miles. Relative to this reality called earth, the smallest unit is the 7α which is defined as the smallest amount of potential energy, mass, or electro-magnetic force perceivable by the assisted (microscope, telescope) or unassisted human eye. If we acknowledge a smallest unit x, that means the reality "builds" itself in a predictable sequence based on x. Whatever x is, it increases sequentially to create *all that is* relative to that reality.

Anything beyond this reality called earth and the layers and interphases above it—atmosphere, Van Allen belt (layer 2)—we call outer space (layer 1). What's it like in outer space?

- It's cold; there is no heat.
- It's dark; there is no light.
- There is no mass and therefore no sound.
- It's a vacuum; there is no pressure and this is one reason why astronauts must wear space suits.

But is it empty? Is outer space the absence of *everything*? Years ago many scientists entertained the concept of an ether (or aether) in outer space. This isn't the same stuff once used as a surgical anesthetic; this outer space ether was thought to be an all-pervading, elastic, massless medium through which light travelled. Early in the 20th century, scientists "proved" to their satisfaction there was no ether and the concept was abandoned.

However, we're going to re-introduce the concept of an ether, but this time in terms of the $_7\alpha$. There are $_7\alpha$ in outer space, lots of them; but they're mostly in total energy, all-linear-motion form. Because there's nothing to get in their way, the deep space $_7\alpha$ can "uncoil" and travel as fast as they want. How fast? They can attain an infinite velocity but most (80%) move at 10^{26} meters per second (m/sec). That's fast! If the universe is 10 billion light-years in diameter, a $_7\alpha$ moving at 10^{26} m/sec can traverse the universe in about one second!

If we accept that all outside this reality is energy, then the entry form into this reality must be energy, and so it is. Relative to this reality (earth), the $_7\alpha$ enters as pure, linear energy. Because linear energy can only exist in a vacuum, the instant it "hits" this reality, it changes. Indeed, it must change if it's to be perceived as part of this reality. Otherwise it merely passes through.

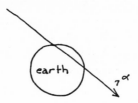

The complete pass-through, however, is so rare as to be inconsequential. What usually happens is that the $_7\alpha$ encounters "something" and

- slows down (loses linear velocity).
- alters its linear course (changes direction).

What we're going to do is to follow a $_7\alpha$ through the layers to the surface of the earth. Remember, we can think of the layers as greater and greater concentrations (density) of $_7\alpha$, with each layer existing at a predictable distance from earth. We number the layers from 1 to 13, and call outer space layer 1 for convenience even though it isn't strictly a layer.

Because layer 2, the Van Allen belt, is the first pure area of perceptual reality and its $_7\alpha$ represent the smallest amount of form humans are capable of perceiving as real, let's introduce our linear, outer space (layer 1) $_7\alpha$ to the Van Allen belt and see what happens. Our $_7\alpha$ has three options as it approaches layer 2:

- It may pass through the layer unchanged.
- It may be deflected by the layer and return to outer space.
- It may enter the layer and pass through, deflect off, or combine with another $_7\alpha$.

Throughout our discussion of the interactions of $_7\alpha$, we will refer to their spin and linear velocities as well as their total velocities. Let's review these parameters once again, and see how they relate to each other.

Because outer space $_7\alpha$ are all linear motion (energy) entities or moieties*, we may designate them numerically as

$$V_S/V_L = 0/100 \qquad V_S = \text{percent of total velocity}$$
$$\text{expressed as spin}$$
$$V_L = \text{percent of total velocity}$$
$$\text{expressed as linear velocity}$$

Keep in mind that

$$V_T = V_S + V_L \text{ and } V_T = \text{total velocity}$$

The use of a ratio to express V_S and V_L is handy because we can have an outer space $_7\alpha$ travelling at 10^{26} m/sec or one moving

---

*A moiety is defined as (1) a half or (2) a part, portion or share of indefinite size. (*American Heritage Dictionary*, Houghton-Mifflin).

at a slower speed, say 10^{15} m/sec and if both moieties are all linear velocity and no spin, they're still expressed as

$$V_S/V_L = 0/100$$

In the case of our faster $_7\alpha$

$$V_T = V_S + V_L$$
$$10^{26} \text{ m/sec} = 0 + 10^{26} \text{ m/sec}$$

and our slower $_7\alpha$

$$V_T = V_S + V_L$$
$$10^{15} \text{ m/sec} = 0 + 10^{15} \text{ m/sec}$$

What's the nature of the $_7\alpha$ within the belt? Although many probable configurations can exist, the predominate form (80% of the moieties) is half linear motion and half spin. Our designation of these $_7\alpha$ is

$$V_S/V_L = 50/50$$

What are their *absolute* values of spin and linear velocity? The average total velocity of a Van Allen $_7\alpha$ is also 10^{26} m/sec, so

$$V_T = V_S + V_L$$
$$10^{26} \text{ m/sec} = (5 \times 10^{25}) + (5 \times 10^{25}) \text{ m/sec*}$$

In other words, half of the total velocity occurs as spin and the other half as linear velocity.

Now let's talk specifically about spin. We measure spin (V_S) in m/sec rather than rpm or rps. We do this so both spin and linear motion will be expressed in the same units. Any rotational rate can be expressed in linear terms as long as we know the diameter of the rotating object. Suppose we have an automobile tire 1 meter in diameter rotating at 1 rps. How fast would the car move? First let's look at a diagram of the tire:

---

*Half of 10^{26} is 5×10^{25}.

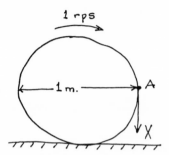

Any point (A) on the circumference has a linear equivalent with a velocity, X, which can be calculated using the formula

$$X = \Pi \, d \, \omega$$

X = linear equivalent of spin
Π = 3.14159 . . .
d = diameter
ω = rotational rate (spin in rps)

So our rotating tire produces a forward (linear) motion of

$$X = \Pi \, (1 \text{ m}) \, (1 \text{ rps}) = 3.14159 \text{ m/sec}$$

or about 7 mph. This is the linear equivalent of its rotational motion, i.e., its velocity of spin expressed in linear terms.

Because we know the linear velocity equivalent of the $_7\alpha$ in the 50/50 Van Allen belt configuration, we can solve for ω (its spin in rps) by turning the formula around:

$$\omega = \frac{X}{\Pi d}$$

We said before that the average of a $_7\alpha$ is 10^{-13} mm (which is the same as 10^{-16} m) so its rotational rate is

$$\omega = \frac{5 \times 10^{25} \text{ m/sec}}{\Pi \, (10^{-16} \text{m})} = 1.6 \times 10^{41} \text{ rps}$$

Therefore, although we are expressing spin in m/sec when we use the notation V_S, it can always be converted to its rotational equivalent.

Now let's look at some characteristics of the basic 50/50 $_7\alpha$ within the layer. A single $_7\alpha$ of half spin, half linear velocity doesn't exhibit true polarity or charge. Polarity is only exhibited when *two* or more $_7\alpha$ are compared because it describes a relationship. If the $_7\alpha$ spin and translate in the same direction, the result is a *monopolar* entity. If two or more $_7\alpha$ spin or translate half in one direction and half in the opposite direction, the resulting $_7\alpha$ function *bipolarly*. Again, if spin and/or linear velocity are in the same direction, the $_7\alpha$ function *monopolarly*. If both spin and linear translation occur in opposite directions, the $_7\alpha$ function bipolarly.

Think of two skaters facing each other and spinning (A) versus two facing opposite directions and spinning (B):

The resulting patterns created by the two pairs look like this:

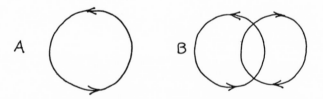

Pattern A functions uniformly (monopolarly) whereas pattern B appears to pull away from itself, i.e., it functions bipolarly.

Now suppose the $_7\alpha$ interact in such a way that their directions of spin and/or linear motion occur at right angles to each other:

These are called *perpendicular* relationships.

Therefore, we may say that within the Van Allen belt (layer 2) the following $_7\alpha$ or combinations of $_7\alpha$ can occur:

- Uncharged *singular* $_7\alpha$ passing through.
- Monopolar combinations of two or more $_7\alpha$ having the same direction of spin and/or linear motion.
- Bipolar combinations of two or more $_7\alpha$ having opposite directions of spin and/or linear motion.
- Perpendicular combinations of two or more $_7\alpha$ whose directions of spin and/or linear motion are perpendicular to each other.

We can represent these four conditions as

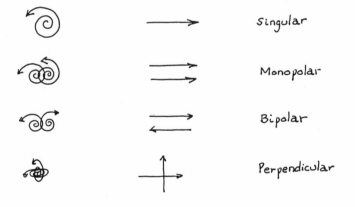

In addition to being affected by the direction of the spin and/ or linear motion of other $_7\alpha$, combinations of $_7\alpha$ are also influenced by the relative total velocities (V_T) of those other $_7\alpha$. Combinations occuring among $_7\alpha$ of different total velocities are so unstable they are referred to as *incoherent*. Monopolar $_7\alpha$ having the same V_T are *coherent;* bipolar $_7\alpha$ of equal V_T form *composites*. When equal velocity $_7\alpha$ combine perpendicularly, the result is an *inherent* combination.

This preliminary information about interactions and polarity is important because much of what we describe within the Van Allen belt holds true for all layers and the earth itself. To be sure, the Van Allen belt has some special qualities because it's truly the interface between outer space and the outer reaches of the earth's aura or atmosphere, but the basic principles still hold.

Let's take a linear $_7\alpha$ from outer space (layer 1) and introduce it into layer 2. Again, like its first encounter with the entire belt, the $_7\alpha$ may

- Pass through the Van Allen $_7\alpha$ unchanged.
- Combine with or be deflected off a Van Allen $_7\alpha$ obliquely.
- Combine with or deflect off a Van Allen $_7\alpha$ perpendicularly.

If you've ever closely observed the game of pool, you know there are an infinite variety of reactions that can occur when a translating and spinning cue ball hits the fixed numbered balls. In our game in layer 2, although the cue ball (our outer space $_7\alpha$) only translates and doesn't spin, all the numbered balls (Van Allen $_7\alpha$) are translating *and* spinning and can hit each other *or* combine. You can imagine the number of probable reactions in a game like that!

Although we realize the linear $_7\alpha$ is most likely to encounter a group of $_7\alpha$, we'll describe these interactions as though they occur between singular entities for easier visualization. Let's begin by examining the pass-through reaction. Like those $_7\alpha$ which penetrate the entire layer unchanged, 10% of those encountering another $_7\alpha$

pass through it unchanged. The linear velocity of these incoming $_7\alpha$ from outer space is so fast, it's as though they aren't there at all.

Another 10% of the linear $_7\alpha$ hit the 50/50 $_7\alpha$ of the Van Allen belt obliquely. Even though the probable reactions are infinite, we can make some general statements regarding oblique interactions:

1. The incoming linear $_7\alpha$ which strikes the Van Allen $_7\alpha$ obliquely may

   - Pass through the Van Allen $_7\alpha$ unchanged if its total velocity exceeds that of the Van Allen $_7\alpha$ (10% of all oblique $_7\alpha$ reactions).
   - Be deflected off the Van Allen $_7\alpha$ if its total velocity is less than that of the Van Allen $_7\alpha$ (10% of all oblique $_7\alpha$ reactions).
   - Combine with the Van Allen $_7\alpha$ if its total velocity equals that of the Van Allen $_7\alpha$, forming monopolar, bipolar, or perpendicular pairs (80% of all oblique $_7\alpha$ reactions).

2. When interactions occur, whether deflection or combination, the *Van Allen* $_7\alpha$ experiences changes in its total velocity and direction of translation and/or spin, but maintains its 50/50 configuration.

3. When interactions occur, whether deflection or combinations, the *linear* $_7\alpha$ loses linearity and picks up spin; i.e., changes its configuration, *and* decreases its total velocity. The direction of its spin is determined by the Van Allen $_7\alpha$.

Let's look at some oblique interactions where an incoming $_7\alpha$ is deflected obliquely off a Van Allen entity, i.e., where its V_T is less than that of the Van Allen $_7\alpha$. The large arrows are used to denote the direction of $_7\alpha$ translation. We already know the Van Allen $_7\alpha$ maintains its 50/50 configuration, but loses total velocity and changes direction. At the same time, the linear $_7\alpha$ decreases its total velocity and picks up spin. Let's see how this works. Suppose we have a totally linear $_7\alpha$ (A) coming in from outer space, and a 50/50 Van Allen $_7\alpha$ (B) translating to the left in our diagram and spinning in a counter-clockwise direction characteristic of the layer:

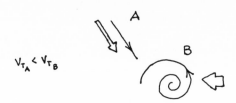

When the two $_7\alpha$ collide obliquely, the Van Allen $_7\alpha$ (B) is knocked toward layer 3, slows down, and reverses direction, now translating to the right (B'). This change in direction is not unlike what happens when a thrown ball encounters a bat.

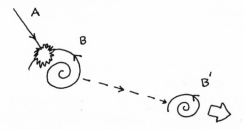

Now let's look at what happens to the linear $_7\alpha$ (A) from outer space. The impact also slows it down and it begins to increase spin (A') in a direction opposite that of the Van Allen $_7\alpha$.

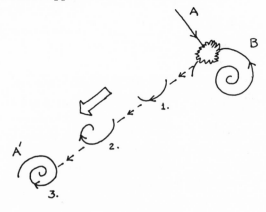

What happens if the linear $_7\alpha$ hits the Van Allen $_7\alpha$ from the opposite side, i.e., from the right in our diagram?

$$V_{T_A} < V_{T_B}$$

In this case the Van Allen $_7\alpha$ (B) still loses total velocity, but this time it's propelled toward the next layer but in the *same* direction as its original path (B'). Think of yourself slowly creeping ahead in your car anticipating a green light and being nudged forward by an even more anxious driver behind you.

When this occurs, the linear $_7\alpha$ (A) again slows down, but this time it picks up spin and translates in the *same* direction (A') as that of the original Van Allen $_7\alpha$ path.

From this we can see that the ultimate translational (linear) and spin direction of both $_7\alpha$ are each determined by the other. The linear $_7\alpha$ from outer space (or the preceding layer) determines the direction of translation and total velocity of the $_7\alpha$ within the layer. The $_7\alpha$ within the layer determines the total velocity, translation and spin direction of the incoming $_7\alpha$.

If these oblique interactions occur *almost* perpendicularly, an interesting phenomenon occurs: the linear motion of the $_7\alpha$ is momentarily stopped. Let's follow this process step by step, beginning with an encounter between a linear $_7\alpha$ from outer space (A) and a 50/50 Van Allen $_7\alpha$ (B) spinning in a counter-clockwise direction and translating to the left:

A

$V_{T_A} < V_{T_B}$

B

When the linear $_7\alpha$ encounters the Van Allen $_7\alpha$, it literally absorbs the Van Allen's linear motion and converts it to spin (A'). This leaves the Van Allen $_7\alpha$ momentarily without any translation motion (B'); it's momentarily stopped in its tracks although it continues to spin.

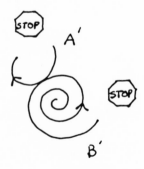

STOP A'

STOP

B'

As soon as the linear $_7\alpha$ begins to spin however, it "rolls" off the Van Allen $_7\alpha$ and propels the latter in a direction toward layer 3 and *opposite* that of its original course (B″). The linear $_7\alpha$ (A″) assumes the original path of the Van Allen $_7\alpha$ within the layer.

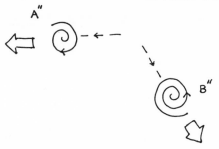

When the linear $_7\alpha$ from outer space (A) encounters a Van Allen $_7\alpha$ (B) near perpendicularly and to the right, the process is reversed.

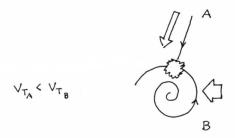

Again, we see the same increased spin imparted to the linear $_7\alpha$ (A′) and inhibition of the Van Allen $_7\alpha$ translational motion (B′).

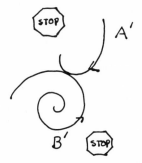

Only this time, when the now spinning $_7\alpha$ from outer space (A″) "rolls" off the Van Allen $_7\alpha$, it propels the latter in the *same* direction (B″) as the Van Allen $_7\alpha$'s original motion.

Thus we may characterize the near perpendicular interactions as follows:

1. There is a period when all translational motion is momentarily stopped.
2. The Van Allen $_7\alpha$ is propelled in the same or opposite direction of translation depending on its original spin and direction of linear motion, and the point of impact.
3. The propelling effect of the near perpendicular interaction is weaker because some of the linear energy has been given up to spin.

The remaining oblique interactions occur between linear and Van Allen $_7\alpha$ having the same total velocity. These interactions form oblique *pairs* rather than passing through each other unchanged or forming changed, but singular (deflected) entities. Let's look at some examples.

If the linear $_7\alpha$ from outer space (A) encounters a 50/50 Van Allen $_7\alpha$ translating to the left and spinning in a counter-clockwise direction (B),

$$V_{T_A} = V_{T_B}$$

the two form a bipolar composite pair, A′ and B′. It is bipolar because the two $_7\alpha$ spin in opposite directions; it's a composite because both $_7\alpha$ have the same total velocity.

If the linear $_7\alpha$ from outer space (A) encounters a Van Allen $_7\alpha$ (B) of equal velocity from the right,

$$V_{T_A} = V_{T_B}$$

they form a monopolar pair (A′ B′); i.e., the two $_7\alpha$ spin in the same direction.

We can summarize the effects of the oblique interactions of the incoming $_7\alpha$ on the $_7\alpha$ within the layer as follows:

- The incoming linear $_7\alpha$ loses velocity and picks up spin, thereby changing its *form* or configuration.
- The $_7\alpha$ within the layer primarily experiences changes in its direction and total velocity, but *not* its form.

Now let's talk about the most common form of interaction within the layer, the head-on or perpendicular reaction. When an incoming, all-linear energy velocity $_7\alpha$ from outer space strikes the 50/50 $_7\alpha$ perpendicularly within the layer, what are the possible results? Again the same three interactions exist depending on the total velocity of the incoming $_7\alpha$ relative to the one within the layer:

1. The linear $_7\alpha$ may pass through the Van Allen $_7\alpha$ unchanged.
2. It may hit the Van Allen $_7\alpha$ and be deflected.
3. It may combine with the Van Allen $_7\alpha$.

As with our other interactions, if the velocity of the entering $_7\alpha$ from outer space is sufficiently great, it may pass through the Van Allen $_7\alpha$ without exerting any effect. These account for 10% of the perpendicular interactions.

Another 10% of the incoming $_7\alpha$ from outer space lack sufficient energy (total velocity) to either pass through or combine, and bump the Van Allen $_7\alpha$ (B) in a path having the same translational direction as the incoming $_7\alpha$. As in all deflection interactions, the linear $_7\alpha$ loses total velocity and picks up spin in the process, with the majority (again 80%) of those so deflected eventually achieving the 50/50 configuration characteristic of the layer.

The remaining 80% of the linear $_7\alpha$ from outer space that hit Van Allen $_7\alpha$ perpendicularly combine and form pairs. Although all combinations of spin and linear motion are possible, the most common is again the 50/50 pair. Such a pair composed of two inherent (equal total velocity) perpendicular 50/50 entities has a special name; it's called a photon. Eighty percent of all the pairs formed in the Van Allen belt are photons.

In this chapter we've discussed the basic options available to any $_7\alpha$ any time it encounters another:

- If its total velocity is greater, it passes through the other $_7\alpha$ unchanged.
- If its total velocity is less, it's deflected, slows down, and changes form (i.e., loses total velocity and picks up spin).
- If the $_7\alpha$ interacts with one of equal velocity, they combine forming a pair whose alignment is determined by the plane of impact. If the incoming $_7\alpha$ enters obliquely, mono- or

bipolar pairs are formed; if they interact perpendicularly, a photon is formed.

In addition to describing these interactions, we also introduced some new terminology. If two or more ₇α exhibit spin and/or linear motion in the same direction, they function *monopolarly*; if their directions of spin and/or linear motion are different, they function *bipolarly*. If two or more monopolar ₇α also have the same total velocity, they are *coherent*. If two or more bipolar ₇α have the same total velocity, they function as *composites*. If equal velocity ₇α align themselves perpendicularly, they form an *inherent* combination. Inherent combinations of two 50/50 ₇α are called *photons*.

Putting it altogether, we can see how the deflection reactions serve to draw ₇α from one layer into another. Every time our linear ₇α deflects off a 50/50 Van Allen ₇α, it assumes the configuration and (ultimately) translational direction of the latter. Meanwhile, the Van Allen ₇α is deflected toward the periphery of the belt where it most likely will encounter another ₇α and change again. Thus, we may view the deflection reactions as the replenishers and stabilizers of the total layer and interphase populations.

The monopolar and bipolar pairs act as charged moieties ready to combine with other ₇α and form even larger groups and configurations. Although the ₇α in a monopolar group are uncharged relative to each other, they function as charged relative to a monopolar, bipolar, or singular ₇α exhibiting a different orientation (direction). The bipolar pairs are quite versatile in that they may function either positively or negatively relative to other single ₇α or ₇α groups.

The inherent perpendicular pairs are the basic energy configuration in this reality. It is from this configuration that all potential, heat, light, sound and pressure originate. So important is this particular concept, we'll spend much more time discussing it. Because the inherent perpendicular pairs are multiplanar, i.e., exist in more than one plane, we need to take a look at the nature of the multiplanar (multidimensional) realm first.

6 | The Multidimensional Realm

In some respects this chapter is a slight detour in our journey towards earth, but it's necessary if we're to understand the life and times of the $_7\alpha$. Before we begin discussing the formation of potential and light, we need to make a side trip into multidimensionality.

If we examine a rectangular solid, almost everyone would agree it has three dimensions: length (l), width (w), and depth (d):

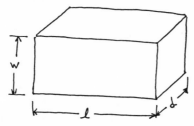

If our block sits on a table in front of us, we would agree it is "here". If someone took it out of the room, then it's no longer here; it's "there", and our "there" could be anywhere—the next room, the next town or the moon. Imagine you're sitting in your room and a friend walks through carrying the block. She enters by one door and leaves by another. Let's also imagine it took her ten seconds to cross the room. To you the block came from one place (or plane of reality), existed in your room (reality) for ten seconds, and went to another plane of reality. Now the block has three new dimensions:

1. The dimension from which it came which may be totally unknown to you (your friend may have flown in from New Delhi while you were sitting in your room in Akron, Ohio).

2. The dimension of time. In your "space-time system" this is ten seconds.
3. The dimension to which the block went—again unknown to you.

For you and your "room reality" the block has a space-time diagram like this:

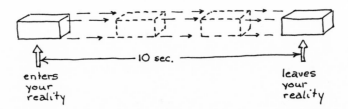

enters
your
reality

10 sec.

leaves
your
reality

Therefore we can say the block is multidimensional or multiplanar, meaning it has planes of reality other than those we usually experience with our physical senses.

We can also look at multidimensionality another way. Suppose you're sitting in a room unable to look anywhere but straight ahead, and your friend carries the same block through the room in the same 10 second span. This time, because you can't follow the block from the instant it enters until it leaves, it appears to have more detail, more form and "realness", when it's directly in your line of vision, but less to either side:

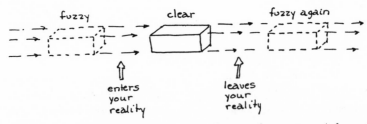

fuzzy   clear   fuzzy again

enters
your
reality

leaves
your
reality

In this case the block still has two unknown dimensions (where it came from and where it's going), but it's time dimension relative to you is much shorter: i.e., the block is totally visible and real only for that part of its ten-second transit time when it's in your direct line of vision.

Therefore, we can say that what we experience as "here" with our physical senses creates *our* plane of reality, *our* singular world. Because of its singular nature we can also refer to it as unidimensional or *uniplanar* when compared to all the infinite probabilities inherent in a definition of "there". For example, when we're sitting in *our* chairs in *our* rooms in *our* houses, we have a very definite image and idea where we are. However, our ideas and definitions of everything else may be quite vague; can you image New Delhi at this time of year? What does it look like in the basement of a tenement on Main Street in Upper Sweatsock, Ohio? We tend to have a clear picture of our immediate reality and lump all others together, even though we may accept that uniqueness and clarity of those other realities exists to their creators.

The important thing is not to confuse the concept of uniplanar reality with the mathematical definition. "Uniplanar" in mathematics means having only length and width, but no depth; i.e., being two-dimensional. Uniplanar realities definitely have depth; how much depth one's reality has is determined by each individual's perceptions.

We experience multidimensionality when we day-dream, sleep, faint, become unconscious.* Our bodies may be "here", but where is the rest of us? Rather than answer that directly, for now we'll just say we're "there". (When someone says of a friend, "He's not all here," there's more truth to that than fiction!)

All conscious-kind is unidimensional and multidimensional, and everything has consciousness. Therefore, everything is unidimensional and multidimensional—you, another person, a rabbit, a fern, a pebble, a cell, a molecule, an atom, and of course, a $_7\alpha$. Keep in mind there's an infinite variety of consciousness. A grain of sand, for example, doesn't worry about its checkbook balance and

---

*Naturally the obvious next step is death which we'll discuss later.

it's unlikely a tree contemplates the Carnot heat cycle of internal combustion. It is fascinating to note, however, that as soon as a group of monkeys on a remote island began washing their yams before eating them, another band on another island began the same practice—no, some monkeys from the first island didn't swim to the second one, and a zoologist didn't transfer monkeys. It just "happened."

A number of years ago an Irish singer by the name of Lonnie Donnegan recorded a song called "The Rock Island Line". In the song he related the story of a train engineer who crossed a bridge and had to pay a tax based upon his cargo. He told the tax collector he was carrying cows, pigs, and sheep; the publican told him he owed no tax on livestock. As our mythical engineer built up speed, he called back to the man on the bridge that he'd fooled him because what he had was a load of pig iron not pigs.

Well we kind of fooled you, too. We said you may look at the $_7\alpha$ as

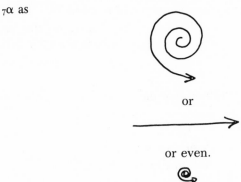

or

or even.

But that's not the whole story. If we have a spiral such as,

we can stretch it into three dimensions by "pulling" on it until it looks like this:

Think of stretching out a Slinky with smaller "loops" at one end and larger ones at the other.

If we stretch a second spiral which is the mirror image of the first,

and put them together, we have:

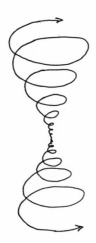

Now let's bend the entire structure around and join the ends:

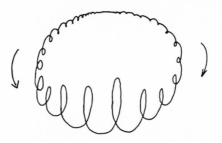

This is the multidimensional form of the $_7\alpha$.

However, before we begin dissecting our multidimensional "spindle", let's talk some more about the concept of multidimensional (multiplanar) vs. unidimensional (uniplanar). Any $_7\alpha$ may traverse realities at will. When we speak of *this reality* we normally mean what takes place on earth and its nearby environs. Because we've been to the moon, many people include that small, craggy planet as a part of this reality. So the term *uniplanar* refers to a single reality or plane which we can represent relative to all other probable realities as:

Is it any wonder some ancients viewed the earth as flat? Or that others talk of experiencing a "slice" of life? If we're in a room without windows, our reality or uniplanarity consists only of what is contained within those four walls, the floor, and the ceiling.

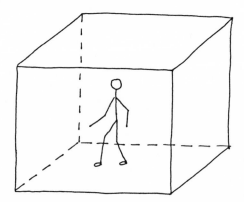

It consists of only that which we can see and experience within that room. Everyone's reality is different, of course, because each person is unique and perceives and experiences things in his or her own way. So when we use the term *reality*, it's a relative term that depends on a reference point defined by one or more of those who share an environment.

The term multiplanar defines multiple (more than one) realities. Because most of us define our reality (i.e., our uniplanar world) in terms of the earth "plane", multiplanarity becomes those planes beyond and outside the earth—deep space, adjacent universes, "time warps", spiritual (energy–essence) forms and the like. Thus we may graphically represent multiplanarity as:

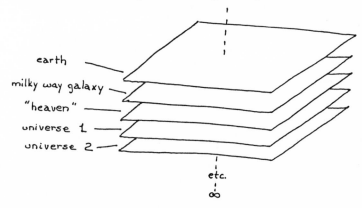

There are an infinite (∞) number and variety of universes and they're all connected via what we can think of as a multiplanar "web". Energy ($_7\alpha$) transfers are made through an interconnecting system of white holes (energy in) and black holes (energy out).

What could be in these infinite universes? Imagine a universe identical to this one, right "next door", except that in "your" copy of this book in that reality the word *your* just used isn't placed in quotation marks. In another universe, the type on these pages is white and the background black. In a third universe, people eat with their feet. In a fourth, the leaves stay on the trees in winter. Now imagine that every single thing you can think of has an infinite number of variations, and you can begin to appreciate what we mean when we speak of the infinite.

The $_7\alpha$ traverses realities at will—it is multiplanar. Its state in this reality is different from its state in other realities. We measure velocity in meters per second, miles per hour, millimeters per millisecond. Another reality might express velocity in quarts per cubic pound, rats per Pied Piper, or green circles per yellow watt. Lewis Carroll's marvelous *Alice in Wonderland* and *Through the Looking Glass* were attempts to explain the UFT in imaginative, non-mathematical ways. As Alice was able to cross the inter-reality zone with no (or little) difficulty, so the $_7\alpha$ has the same ability. Dorothy, in the *Wizard of Oz* was also able to do this; but like Alice in Wonderland, Dorothy wasn't the same being once the transition was made. Both had to learn a new set of rules; but both adapted quickly.

A prevalent opinion today is that this reality is all there is. People believe they have a beginning in this reality, called birth, and at some time have an end: death. Nothing could be farther from what is. Even people who believe there's a beginning in this reality have difficulty deciding when that beginning occurs—at the time of fertilization of the egg by the sperm? Ninety days following conception? At a certain developmental level of the fetus? At physical birth? There's also controversy regarding when death occurs: when the heart stops? When the theta waves of the EEG are quiescent? Thirty minutes after all measurable brain activity ceases? The answer is that there is no beginning or end for you or the $_7\alpha$. In other words, the reality of the $_7\alpha$ in this plane is not

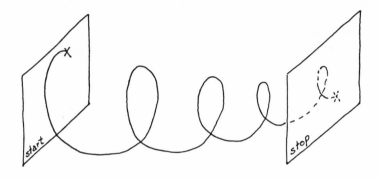

but more like

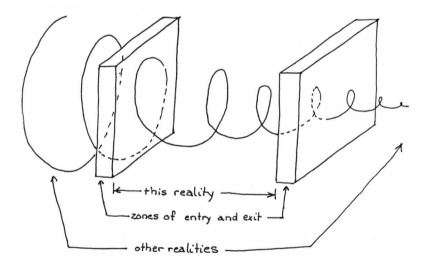

In the previous drawings we show only a portion of the multi-dimensional spiral for simplicity. At other times we'll return to the simpler spiral

because it's easier to visualize.

Earlier in the book we talked about the phenomena of the magic disappearing rock:

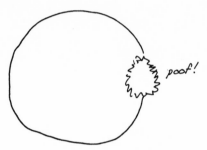

We made the point that physicists often see subatomic particles disappear before their eyes. When these *reappear* like a new particle on the scene, many scientists don't believe they're seeing the "old" particle reappearing at a new point or in a different form.

For example, suppose you were able to follow a flower pot moving on a circular conveyor belt. As time went on you saw a plant begin to grow, flower, produce fruit and die. Now suppose you couldn't see the entire process, but only one part of it from a fixed position for a fixed time. Depending on where you placed yourself, you might see

1. nothing
2. a pot filled with dirt
3. a young plant
4. a flowering plant
5. a fruit-producing plant
6. a dead plant.

Unless you broadened your range of perception, you could believe that the flowering plants were singular entities that appeared and disappeared at the same time and place. You could even use this as evidence or proof that they were distinctly different from the dirt-filled pots your colleague saw from his or her particular vantage point. Such is the dilemma facing many scientists. However, the $_7\alpha$ is able to traverse realities *at will* and time, as we define it here on our planet, is essentially meaningless to any $_7\alpha$ worth its spin.

What defines the particulate nature of mass in this reality is the photon—the unit of light. Unless a particle is "seen", it doesn't exist. As for where the particle "lives" before it exists here and after it "decays", there can only be one answer because all things, sub-nuclear/subatomic particles included, are part of a continuum. If we look for them, we invariably see them as they pass through our reality; if we don't look for them, we don't see them; however, they still exist.

Something that *is*, is either here or somewhere else. That's the definition of *being*. Too often we assume that particles are always here but we're missing them. That's not the case; it's just as likely that the particles are somewhere else. The beautiful thing about $_7\alpha$-based rotational physics is that nothing is ever lost. Some part of everything is always in *all* possible reality planes, including an infinite number of these multi-planes. Think again of the $_7\alpha$ as the original spiral of infinite diameter *and* depth:

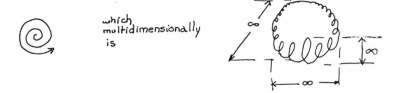

(The symbol, ∞, has been used to denote "infinity.")

Each loop or ring of the $_7\alpha$ represents a probable plane of reality,

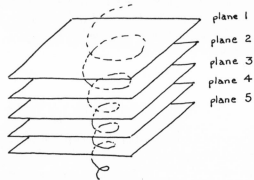

plane 1
plane 2
plane 3
plane 4
plane 5

and each $_7\alpha$ spiral may intersect with the infinite number of planes in any one of an infinite number of ways:

It is important to note that in any of these manifestations of the $_7\alpha$,

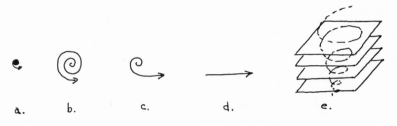

a.　　b.　　c.　　d.　　e.

its identity, the knowledge of what it is where, is known to its entire range of form and energy. All planes intersecting the open $_7\alpha$ spiral in "e" above have access to the entire $_7\alpha$ spiral potential and form. A $_7\alpha$ in all-linear-velocity form ("d") of 8,000 meters per second carries with it all the properties of the $_7\alpha$ at "a" having 10^{-2} (.01) meters per second linear velocity and 10^8 meters per second spin velocity. So although the form of the $_7\alpha$ may vary greatly, it's always at one with itself and always knows what it is and what it can be. The $_7\alpha$ can never be more or less than it is.

Think again of a middle-aged, music-loving, accountant, father, week-end gardener. These are all different forms of the same individual; he is always aware of his *potential* to be any one or all of those things. He is no less of a being when he functions as a gardener, nor more of one when he functions as an accountant or father.

So we can say the "big bang" theorists who hypothesize an extremely dense particle that "ruptured" to form the universe are correct in a way; their error is in thinking it was a particle of the past when in fact it *is* part of every *living* thing. The $_7\alpha$ by virtue of its very being confers life on all things. If there is no time, there is no death or even birth *per se;* all is life. The concept of animate and inanimate is a time-based one. Because of our human sense or perception of time, we arbitrarily decide a rock's metabolism is so slow compared to that of a tree that the rock is "dead". And yet the stone appears a virtual hotbed of activity when compared to the inactivity of certain "dead" planets. Thus death is the most relative concept of all.

In the next chapter we will discuss the form of the $_7\alpha$ in more detail.

7 | The Full Form of the Sub-Seven Alpha

In the last chapter we represented the multidimensional $_7\alpha$ as

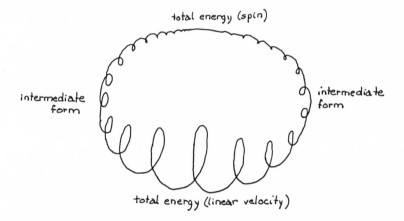

Relative to this perceptual reality, the $_7\alpha$ has two total energy forms: all linear motion (0/100) and all spin (100/0). For us on this planet, the 0/100 form is academic because its average velocity (10^{26} m/sec) is beyond normal human comprehension. The half-lives of these forms are so short as to be meaningless to a reality that (at least presently) relates most physical phenomena, like the speed of switching in computers, to the speed of light (3×10^8 m/sec). This is nothing new. Physicists have found substances with half-lives measured in trillionths of a second yet they are able to build theories based on these extremely transient entities.

Now let's examine our $_7\alpha$ more closely. To begin, let's look at a single loop, P, of our multidimensional spindle, and a small segment, (A), of that loop:

Now let's enlarge segment (A) until it looks like a straight line:

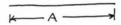

However, always remember that it's actually curved.

Remember our discussion of probability mists in chapter three? We said that any line drawings of the $_7\alpha$ are more correctly made up of dots. Each dot represents a probable location of the $_7\alpha$. The more dots, the thicker and darker the "line"; i.e., the more likely it is the $_7\alpha$ will assume that configuration. Using our probability mist, the perceived straight-line portion of the greater rotation P looks like this:

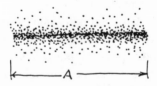

If we return A to P and represent the entire loop as a mist, we see:

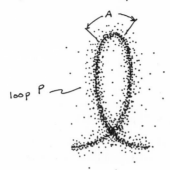

However, if we look very carefully at the pattern created by those lesser probabilities, we see our familiar spiral mist:

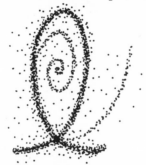

That is, although the majority of the probabilities form a loop, the remainder form a spiral around that loop.

If we pass a plane, X, through our loop and look at it in three dimensions, we see:

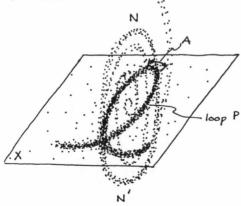

And once again our 80-10-10 rule applies:

1. There's an 80% probability a linear $_7\alpha$ will manifest itself linearly within plane X and loop P.
2. There's a 10% chance it will manifest itself non-linearly; that is, within plane X but *not* in loop P.
3. There's a 10% chance it will manifest non-linearly some-where else—the areas N and N' above and below plane X.

Now let's take a point, B, along our linear path

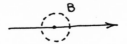

and magnify it. We discover that point B is itself made up of an infinite number of points

and we note that the points are more or less uniformly distributed, statistically speaking.

Let's now pass a vertical plane, W, through B:

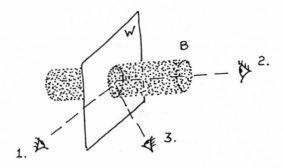

What does an observer see? It depends on where the observer is. An observer at 1 sees a line of dots,

an observer at 2 sees a circle of dots,

and an observer 3 sees an ellipse:

Therefore "point" B has many perceptual forms depending on one's "point" of view.

To reiterate, the linear $_7\alpha$ creates a continuous spiral mist where 80% of the probabilities assume the circumferential or loop form. Of the remaining 20%, 10% occur within the same plane as the linear (circumferential) probabilities and the remaining 10% in all other planes (probabilities). Because of the great space over which the latter function (i.e., 10% of the probable population dispersed over an infinite area) relative to perceived reality, they may be considered nonexistent.

So the *total* form of the $_7\alpha$ is extremely difficult to draw because of all the probabilities. If we use lines to represent the greatest concentration of probabilities, the multidimensional spindle looks like this

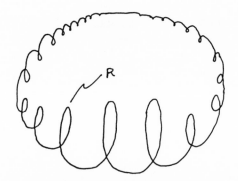

and if we look at any "loop" R, it appears

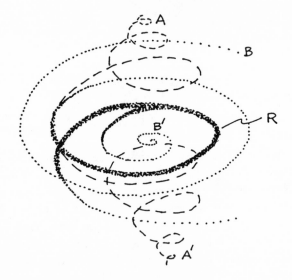

where spiral B–B′ represents the 10% of the probabilities that exist in the same plane as R, and spiral A–A′ represents the remaining 10% of the probabilities in all other planes.

Now let's begin talking about the various intersections (combinations) of $_7\alpha$. We'll use simple, two-dimensional representations such as

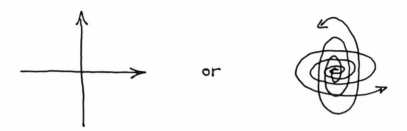

for our $_7\alpha$, but always keep in mind that the actual form is the much more complex multidimensional spiral mist.

In closing, pause a few moments to contemplate what this chapter means in terms of you. Because each and every $_7\alpha$ can be thought of as an infinite set of points, then each and every $_7\alpha$ can be "coded" with all the knowledge relative to any reality in which it finds itself. So, if all can be known to the least, then all can be known to each of us as well. This means there are no secrets, nothing is ever lost, no one or anything is insignificant.

8 | Linear Interactions

In chapter four we described how one $_7\alpha$ encountering another may

1. Pass through it unchanged,
2. Be deflected off it, or
3. Combine with other $_7\alpha$.

In this chapter we're going to begin studying how some of the $_7\alpha$ combinations or pairs form linear potential (i.e., all linear motion).

Remember, the factors determining *how* a $_7\alpha$ functions are its

- Spin.
- Linear velocity.
- Relationship to other $_7\alpha$.

Let's observe *two* $_7\alpha$ in outer space and see what can happen to them. What kinds of combinations can occur between two $_7\alpha$? From chapter five we know they can form

1. Monopolar pairs in which both $_7\alpha$ have the same direction in the same plane.

   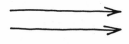

2. Bipolar pairs in which the $_7\alpha$ have different directions in the same plane.

3. Perpendicular pairs in which the $_7\alpha$ move in directions that are at right angles to each other; i.e., they move in different planes.

When we speak of relationships in the energy states—potential, light, heat, sound, pressure—all these forms are variations of the same configuration, the perpendicular pair. However, we're speaking of a very special type of perpendicular pair, the *bisecting* perpendicular pair. What makes a bisecting pair

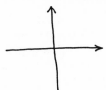

different from a non-bisecting pair?

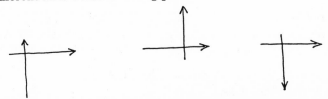

As you can see, the $_7\alpha$ forming a bisecting perpendicular pair are always equal and opposite. There's always as much potential on one "side" of the intersection as the other.

In outer space the most common form is the bisecting perpendicular linear $_7\alpha$ pair. Doesn't it seem that, in such an infinite region as outer space, the probability of a bisecting intersection would be close to zero? After all, any of an infinite number of points along either of the linear $_7\alpha$ could be the point of intersection. So why is the perpendicular bisection the prevalent form?

In order to answer these questions we must expand the concept of mono- and bipolarity described in chapter five. In that chapter we said mono- and bipolar were terms used to describe relationships *between* two or more $_7\alpha$. But what is the relationship that exists *within* a $_7\alpha$ as it moves linearly and/or as it spins? (Because the $_7\alpha$ in outer space are linear, we'll use linear examples. However, the same holds true for those exhibiting spin. In the case of spin we merely replace linear directional words such as east, west, north, south, etc., with clockwise and counterclockwise.)

Let's consider an object moving linearly from left to right or west to east:

$$W \longrightarrow E$$

We may also compare our west to east motion to charge,

$$\underset{(-)}{W} \longrightarrow \underset{(+)}{E}$$

calling east positive to west's negative because the system is moving toward $(+)$ E and away $(-)$ from W. This change or motion *within* a system is called its *intrinsic charge*. Intrinsic charge tells us in which direction the entity is moving relative to where it was before.

If we have *two* moving systems within the same environment, they also exhibit charge relative to each other. If both move in the same direction at the same speed, they have the same *extrinsic* charges:

$$\ominus \quad \underset{(-)}{W_1} \longrightarrow \underset{(+)}{E_1} \quad \oplus$$

$$\ominus \quad \underset{(-)}{W_2} \longrightarrow \underset{(+)}{E_2} \quad \oplus$$

If they move in opposite directions, they have opposite *extrinsic* charges:

What happens if system 1 moves in the same direction but at a *faster* rate than system 2?

Because velocity is also a function of direction like charge, as velocity increases, charge increases proportionately. Thus system 1 functions more positively (with greater *extrinsic* positive charge) than system 2. Similarly system 2 functions less positively, i.e., negatively, relative to system 1.

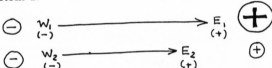

Think of two women walking down a road together. Each has her own method of getting from one place to another, her intrinsic "charge". In addition, both women move in relationship to each other; i.e., extrinsically. The singular pedestrian measures her pace in terms of her own internal standards, whereas in the company of a friend she's more likely to match her pace to her companion. If her friend increases her pace and our walker is unable to keep up, although they will still be going in the same direction there will be more distance/difference between them.

In other words, charge is a function of *differences*. The differences in direction occurring within a single entity are its intrinsic charge(s). Those differences which occur between two or more $_7\alpha$ are their extrinsic charge(s).

Now let's expand some other terms from chapter five. If the intrinsic charges of two $_7\alpha$ are the same

the two entities are *coherent*. If their intrinsic charges are opposite,

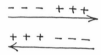

they are *composite*.

As you can see, by definition all coherents are monopolars and all composites, bipolar. However, there can be entities forming both mono- and bipolar pairs which are neither coherent nor composite; their extrinsic charges are in the same direction but their intrinsic charges are not aligned (their total velocities are not equal). For example, our two walkers can take the same number of steps, breaths, burn off the same number of calories to cover exactly the same distance. If they walk in exactly the same way to cover the same distance in the same direction, they function monopolarly and coherently. If they walk in exactly the same way and cover the same distance but in opposite directions, they function bipolarly and compositely. If they cover the same distance in the the same direction but pace themselves differently, they function monopolarly but *not* coherently. Similarly, if they cover equal but opposite distances using dissimilar walking styles (rates), they function bipolarly but not compositely.

This is often a difficult concept to picture at first because we tend to see charge as a *thing* rather than a difference or direction.

(We've all gotten a "jolt" from an electrical outlet and use phrases like "I got a charge out of that.") Simply remind yourself that east is *different* from west, north *different* from south, fast *different* from slow, and you'll quickly become accustomed to this way of thinking. Again, we use charge as an indicator and measure of *direction and amount* of motion.

Now let's consider our two approaching linear moieties A and B in outer space. If we have two infinitely long linear $_7\alpha$ approaching each other, it would seem that the most likely combination would be non-bisecting, but we know that isn't the case. Let's look at some examples and see what happens. First, let's look at two linear $_7\alpha$ approaching each other at some non-bisecting point where $V_{T_A} > V_{T_B}$; i.e., the extrinsic charge of A is greater than B:

Because B's lesser velocity functions as a relative negative to A's more positive charge, B is pulled toward A.

However, the resulting pair is most unstable. Although the two $_7\alpha$ are now moving in the same direction (i.e., have the same extrinsic charge), most of their intrinsic charges repel each other. As you can see, the two $_7\alpha$ are held together by a most weak bond at X. Because of this, such combinations have half lives so short as to be non-existent relative to this reality.

What happens if B's extrinsic charge functions as positive to A's ($V_{T_A} < V_{T_B}$)?

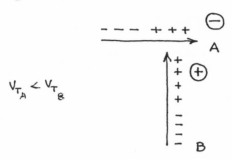

This time, because, of the relative differences in extrinsic charges, A is drawn toward B:

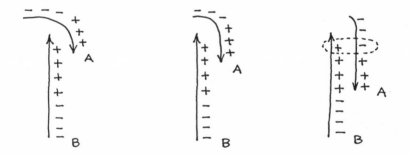

The reaction continues until the opposite intrinsic charges of A and B coincide, again forming a most weak bond. However, whereas our previous combination resulted in a most unstable monopolar pair, this interaction yields an equally unstable bipolar pair. In both cases the misalignment of the intrinsic charges is insufficient to counter the nonaligned extrinsic forces and the unions are quite ephemeral.

If we intersect our $_7\alpha$ at the intrinsically negative end,

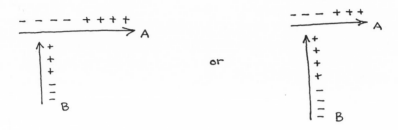

mirror images of the same unstable combinations are formed.

Now, what happens when $V_{T_A} = V_{T_B}$ or $_7\alpha$ A and $_7\alpha$ B have the same extrinsic charge? Before we explore this combination, we have to take a closer look at the nature of intrinsic charge.

It's pretty obvious that if a linear potential is intrinsically positive at one end and negative at the other it must have a neutral area, an area (A') where positivity gives way to negativity and vice versa:

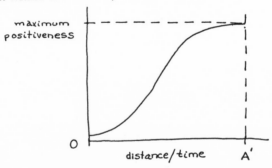

This shift in A' isn't sudden, but rather occurs gradually. If we plot the shift within the area A', we see

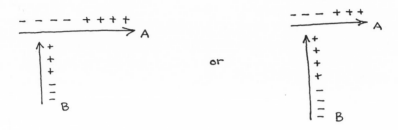

This neutral zone is instrumental in the formation of perpendicular pairs.

If the potentially non-bisecting $_7\alpha$ approach each other at the same velocity, one of two combinations may occur. Because their extrinsic charges (total velocities) are the same, if and where they combine is determined by the intrinsic charges following the basic rule of magnetics: likes repel, opposites attract.

If $V_{T_A} = V_{T_B}$ and the two $_7\alpha$ intercept at their opposite ends (that is, A's intrinsic positive meets B's intrinsic negative),

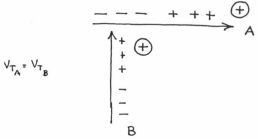

a most unstable perpendicular combination is formed.

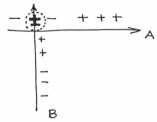

Because some of B's motion has been lessened via its interaction with the negative intrinsic charge of A, it's now moving more slowly and therefore is less extrinsically charged than A. It's unstable because while it's being held by this single intrinsic bond, the *now unequal extrinsic* charges begin to exert their influence.

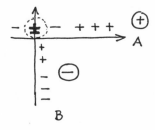

Thus A tends to pull the negative end of B toward it, straining and eventually breaking the weak intrinsic bond.

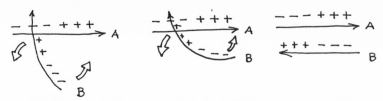

When the bond breaks, A and B immediately form a stable bipolar composite, a pair having opposite alignment of both intrinsic and extrinsic charges, and the same total velocity.

If the nonbisecting 7α approach each other at the same velocity (i.e., having the same extrinsic charge) and in the region of like intrinsic charge, the neutral area comes into play.

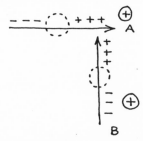

Because the extrinsic charges are equal, the fate of the pair is again determined by the intrinsic charges. Here the basic rules of magnetics (likes attract, opposites repel) once again apply and the more positive "head" of B is repelled toward the more negative end of A:

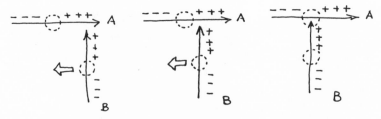

When B reaches the neutral zone of A, it slides easily through

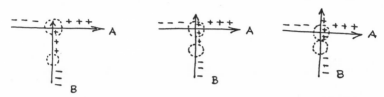

with the intrinsic positive and negative charges of A serving to maintain B's path within the neutral zone by simultaneously attracting and repelling it. Once the neutral zone (the "center") of B coincides with A's,

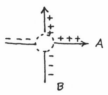

a most stable bond and unit is created. Because of its bisecting, equal and perpendicular qualities, this combination is called an *inherent* pair.

We may summarize the most common results of non-bisecting perpendicular interactions as follows:

1. If the extrinsic charges of the two $_7\alpha$ are unequal, they form highly unstable mono- or bipolar pairs.
2. If their extrinsic charges are equal, they may
   a. initially form highly unstable perpendicular combinations which break down into stable composite bipolar pairs.
   b. form highly stable inherent pairs.

We know from magnetics that equals and opposites strongly attract. Now let's see what happens when our linear-potential, perpendicularly-aligned $_7\alpha$ approach each other at a *neutral zone*. Although these reactions are quite similar to the ones we've just discussed there are some important differences, too. First, let's bring two $_7\alpha$, A and B, together where the $V_{T_A} > V_{T_B}$ or A's extrinsic charge is greater than B's:

Simple magnetic theory again tells us that as the positive end of B approaches A, they should repel each other *if* their charges are the same. However, even though the positive intrinsic charges of A should repel those of B, because of A's greater total velocity B functions extrinsically negative to A's positive and is pulled towards it.

In this case the two $_7\alpha$ form a monopolar pair, one where both components move in the same direction. Thus the *intrinsic* charges of A attract the weaker charges of B on a one-to-one basis, and B's functionally negative extrinsic charge relative to A serves to hold the aligned $_7\alpha$ weakly together, forming a most unstable pair.

If the total velocity and charge of $_7\alpha$ B is greater than $_7\alpha$ A ($V_{T_A} < V_{T_B}$), a different combination is formed. Again the entity with the slower velocity functions negatively,

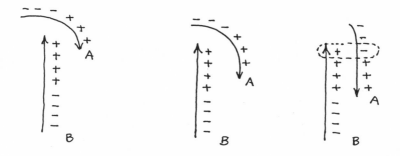

$$V_{T_A} \angle V_{T_B}$$

and is pulled toward the other:

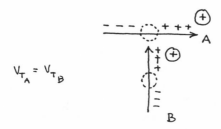

This results in a bipolar pair because each component maintains its opposite extrinsic charge. Like all other combinations involving $_7\alpha$ of different total velocities, the pair is quite unstable.

If $V_{T_A} = V_{T_B}$, the two $_7\alpha$ easily combine

$$V_{T_A} = V_{T_B}$$

to form a highly stable inherent pair,

with B easily sliding through the neutral area as previously described.

The combinations resulting when one $_7\alpha$ aproaches the neutral zone of another are

1. If $V_{T_A} > V_{T_B}$, an unstable monopolar pair is formed.
2. If $V_{T_A} < V_{T_B}$, an unstable bipolar pair is formed.
3. If $V_{T_A} = V_{T_B}$, a stable inherent pair is formed.

We can now see how both non-bisecting and bisecting interactions can produce stable inherent pairs. However, that doesn't explain why the majority of the $_7\alpha$ interactions in outer space produce inherent pairs. The one thing that governs the formation of mono- and bipolar pairs (whether stable or unstable) versus inherent pairs is total velocity or extrinsic charge. If the total velocities of the intersecting $_7\alpha$ are different, unstable mono- or bipolar pairs are formed. If the total velocities of the intersecting moieties are equal, 50% of the non-bisecting interactions (those occurring between the positive ends of the moieties) produce inherent pairs. All of the interactions between two equal velocity $_7\alpha$ where one $_7\alpha$ is opposite the neutral mid-point of the other result in inherent pairs.

Obviously the more $_7\alpha$ of equal total velocity, the greater the probability of inherent pair formation. So the question is: in outer space, is it more likely the $_7\alpha$ have the same or different total velocities? An accepted "truth" says that moving objects within a vacuum continue moving at the same velocity unless acted on by something else. Therefore, because we know outer space functions as a vacuum, we can logically assume the majority of the $_7\alpha$ have the same velocity until they encounter another.

Because the inherent (bisecting perpendicular) alignment is the most common form in outer space, let's spend a few moments looking for evidence of similar structures in the world around us. Do you recognize this intersection as the standard x-y system of Cartesian coordinates?

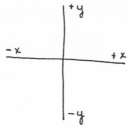

Let's represent this pair multidimensionally by connecting the $_7\alpha$ ends,

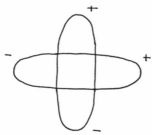

and note its similarity to Bohr's model of the atom. If many $_7\alpha$ combine, we see

and if many $_7\alpha$ pairs of differing velocities combine, we can create "electron shells".

More philosophically, think of two groups of people having opposite views (extrinsic charges) on the same issue, one preferring war, the other peace for example. Within each group there are inevitably those who are strongly, moderately or weakly dove- or hawk-like (intrinsically charged). As they approach each other from their opposite directions, their equally strong forces tend to repel each other blocking any chance of interaction. However, if they maintain their opposite (perpendicular) alignment, they eventually repel each other to a point where one group is aligned with the moderates of the other. When this occurs, the strong and weak doves and hawks are able to interact with each other as *individuals* (intrinsic charges) rather than mutually repulsive and exclusive *groups*. As the interaction continues, the strongly oriented individuals are drawn easily through the moderate environment until their own moderates are drawn into the interaction. At this point a most stable but completely different unit is created. Neither side has given up any of its "power", no single individual has lost his or her identity. However, whereas the original two groups saw themselves as two equal and opposite forces which could not possibly co-exist (one's survival depended on the destruction of the other), they now realize a mutual recognition of the strengths and weaknesses of both orientations. This enables them not only to co-exist but also co-create something much stronger and more stable than that inherent in either one alone.

In closing, we can make the following statements about the nature of bisecting perpendicular (inherent) configurations:

1. They all produce energy.
2. They function both intrinsically and extrinsically.

3. The singular moieties maintain their identity.
4. The pair has its own identity.
5. The pair is very stable.

9 | Let There Be Light

Now that we know how bisecting perpendicular pairs form, let's see how some of those pairs function in our reality. First, let's return to outer space and take another look at our linear, bisecting perpendicular pair,

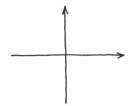

keeping in mind that what appears linear to us is really,

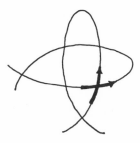

and that even these "loops" are parts of two infinitely large, multidimensional spindles:

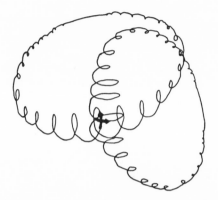

How can we numerically express our linear pair? Because both $_7\alpha$ are all linear velocity ($V_L = 100$) and no spin ($V_S = 0$), we use the now-familiar terminology

$$\frac{V_S}{V_L} = \frac{0}{100}$$

to express each $_7\alpha$ singularly. To designate the $_7\alpha$'s relationship to each other we simply use a fraction:

$$\frac{0}{100} \Big/ \frac{0}{100}$$

This is one of the stable energy forms—pure (linear) potential.

Remember what we said about $_7\alpha$ alignment and intrinsic and extrinsic charge in chapter eight? The more alike the $_7\alpha$ are, the more likely changes will be made in response to intrinsic charge; the more unlike the $_7\alpha$, the more extrinsic charge comes into play. Think of two people speaking the same language (having the same extrinsic charge); they communicate or bond in terms of the same words and phrases (intrinsic charges). However, if one speaks French and the other English, individual words and phrases have no meaning. This "incoherent" pair must find a common ground— drawing pictures, sign language—before effective communication can occur.

Thus in order for a stable energy-producing pair to form, two ...itions must be met:

1. Orientation—they must bisect perpendicularly.
2. Total velocity—they must be equal.

...ther words, the $_7\alpha$ must form an inherent pair.

As we know, these two conditions are concurrent. Perpendic-
...bisection can *only* occur when the potentials (total velocities)
...e intersecting $_7\alpha$ are equal. So all bisecting perpendicular pairs
... energy, some more "pure" than others. If the pair

$$\frac{0}{100}\Big/\frac{0}{100}$$

is pure potential or linear energy, what's

$$\frac{10}{90}\Big/\frac{10}{90}?$$

If the 0/100 $_7\alpha$ is all linear motion, then 10/90 must have less of its
total velocity (only 90% of it) expressed as linear motion and the
remainder as spin. Remember in chapter five we spoke of the inherent combination of two 50/50 $_7\alpha$ as the photon, the basic unit of
light energy? Because the 10/90 pair of $_7\alpha$ occurs between the pure
linear pair and the photon,

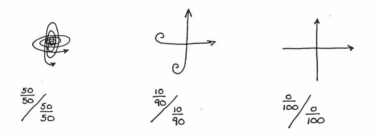

$$\frac{50}{50}\Big/\frac{50}{50} \qquad \frac{10}{90}\Big/\frac{10}{90} \qquad \frac{0}{100}\Big/\frac{0}{100}$$

we can say it represents a form of energy between pure light and
pure linear potential, but exhibits more of the characteristics of the

latter. Similarly, an inherently aligned pair of 40/60 $_7\alpha$ create an energy form more light-like and less like potential.

Because our reality is totally dependent on light, and light is such a familiar energy form, let's examine this form of energy first. In chapter five we learned that the basic photon looks like this:

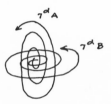

Again we notice it's a bisecting perpendicular pair, so we know

$$V_{T_A} = V_{T_B}$$

and

$$\frac{V_{S_A}}{V_{L_A}} = \frac{V_{S_B}}{V_{L_B}} = \frac{50}{50}$$

From our discussion of the layers in chapter four we know 80% of each layer is made up of $_7\alpha$ combinations and 80% of these (64% of the total) are

$$\frac{50}{50} \bigg/ \frac{50}{50}$$

or photons. Now according to traditional linear physics, the photon is the smallest "particle" visible to the human eye. However, that visible photon is really made up of 10^{13} rotational $_7\alpha$ photons per cubic millimeter (mm^3). That is, in order for the human eye to register "photon", there must be a concentration of 10^{13} perpendicularly aligned 50/50 $_7\alpha$ pairs per cubic millimeter of the visual field. To be sure, we could have called the singular 50/50 pair by a different name; however, because the perceived form has the same characteristics as its smallest component, this seems to complicate things unnecessarily. Think of having different sized prints made of the same photo of your aunt. You still recognize her whether the print is wallet-sized or 8″ x 10″. To be sure, a print may be so small you can't see your aunt's image at all, or so large you can't see the entire image and therefore see only parts of it. But if you're aware of the process, you feel pretty confident the "aunts" in the extreme-sized images beyond your perception are the same as the one(s) in those prints you can easily see.

Because the number of $_7\alpha$ is the same in each layer, the concentration of photons is much less in the more spacious outer layers. Although a $_7\alpha$ may assume the 50/50 perpendicular alignment, it has so much room to spread out, (a) it looks linear and (b) the numbers necessary for perception aren't available. As we proceed towards the earth and the concentration of $_7\alpha$ increases, they no longer have much room to spread out and the necessary numbers per unit area for perception are present.

The function of energy in all forms (potential, light, heat, sound, pressure) is to "produce" mass. The only evidence we have that energy exists is what it does; and what it does is enable us to perceive the changes of one form of mass into another. Photosynthesis causes plants to grow. High heat decomposes mass. Great air pressure can destroy buildings. If it weren't for change, we'd have no need for energy. If it weren't for energy, we'd have no way to perceive mass. Because of this, try as we may, we can't remove the concept of perception from energy. Perception is a function of energy; without energy (light) there's no perception (vision). Without energy (sound), there's no perception (hearing). Without energy (pressure), there's no perception (touch). Without energy (heat), there's no perception (temperature). Without perception there is nothing. Think of the view from a space ship in outer space. Because of the paucity of energy, there is a paucity of change. Because of a lack of change, the environment is inherently empty. Or think of the moon's rather dull lifeless environment. Lacking much of any common energy form, it's void of all but the crudest components.

We know the most common form in outer space is the linear $_7\alpha$ and the photon is the most common form in the Van Allen belt. So how do we get from

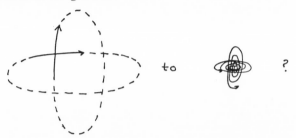

to

?

As we said, it's a matter of concentration; the linear $_7\alpha$ in outer space are only linear relative to here. Like all $_7\alpha$ they conform to the infinitely large spiral mist. However, two things occur when two $_7\alpha$ form a perpendicular pair:

1. The $_7\alpha$ hold each other together.
2. They are drawn to other $_7\alpha$.

Because the perpendicular pair functions as positive *and/or* negative (*not* neutrally), it can go in any direction and is attracted to a broad range of $_7\alpha$ configurations in addition to the singular $_7\alpha$, or other $_7\alpha$ pairs. Furthermore because of its infinite size in outer space, we can think of its perpendicular probabilities as delicate sensors for other potential ($_7\alpha$) activity. Think of an infinitely large spider's web composed of an infinite number of sensitive threads. Any movement on the threads is immediately recognized; because the threads are infinite, *all* change or motion is instantly recognized. In such a way, then, the linear pair draws toward the "flies", the $_7\alpha$ of the Van Allen belt.

A single $_7\alpha$ pair can literally sense the entire universe or, more precisely, all that is. For if the circumference of our perceived linear $_7\alpha$ probability is infinite (as it must be because our linear $_7\alpha$ is defined as infinite), the pair must literally encompass all that is.

As concentration increases, the amount of available space decreases proportionately. As the $_7\alpha$ become more concentrated, it becomes more and more difficult for them to manifest linearly, and therefore the *probability* of a $_7\alpha$ manifesting linearly decreases. Put another way, the amount of spin is directly proportional to concentration and inversely proportional to linear motion.

Because of this, we can see how single or paired $_7\alpha$ change form *relative to this reality,* appearing linear in outer space and exhibiting more and more spin as the concentration increases. Merely in response to concentration, our pair appears

A   B

as it moves from outer space earthward.

Don't think of this as the $_7\alpha$ "shrinking" so much as manifesting a different part of its probability mist. Remember our spinning skater? She can spin with her arms wide-spread as well as held tightly against her body. When she exhibits one form, the potential or ability to exhibit the other isn't lost.

As we look at the progression above, we can see that whether or not a bisecting pair manifests as potential or light, merely depends on where it is relative to the perceiver (observer). At end A we "see" linear energy and at B we see light (photons). If the human retina were geared to recognize *any* inherent 50/50 pair as "light" the entire spectrum from A to B would be perceived as light. Those inherent 50/50 pairs which pass through the earth and its atmosphere as unseen energy because of their high V_T would be seen. Of course, if our eyes were so attuned, we'd see nothing else; all would be light.

Think of all the theological references to great lights, halos, transfigurations. In these situations individuals were/are able to shift their perception into the realm of that which isn't normally seen. Their brains register "light" where normally they register "energy" or "nothing." This is merely a different method of seeing into another reality which, while startling and often perceived as a reason for all sorts of uniplanar behavioral and belief changes, isn't nearly so specific or controllable as that available in extending one's *energy* awareness—via the dream or meditative states, for example.

In essence what those who see great lights succeed in doing is bringing "there" here; but because they choose to bring it here in the most common form (based on visual perception), they have insufficient visual capabilities to convert that much photonic energy to meaningful visual images. Think of having a full orchestra playing in a perfectly balanced large hall versus trying to cram all the players into a phone booth. The resulting noise is certainly loud, but hardly representative of that which the orchestra normally creates, nor that which is perceived by the audience in the hall. Although such blinding lights are manifestations of there here, they're not representative of the whole picture.

If the paths of the component $_7\alpha$ in the inherent 50/50 pairs are sufficiently large or small they appear linear or point-like. If they are between these two (non-perceptual) extremes they appear as light. Thus the difference between light and potential is what we see and don't see.

The photon of the Van Allen belt isn't a light measurement because we can't see it. Although scientists refer to the photon (remember their photon is composed of 10^{13} 50/50 inherent $_7\alpha$ pairs per mm^3) as the unit of light, they also define it in terms of energy. It's that amount of energy given off when one electron jumps from one shell (rotation) to another. They also recognize the presence of these photons in the Van Allen belt. Thus, although it's not a unit or packet of light, what's detected in the Van Allen belt is part of the same continuum as light.

Doesn't it seem logical that the photonic configuration manifests itself in forms other than light just as all things manifest in different forms? We can look at this phenomenon of multiple forms several ways. We may refer to the 50/50-50/50 configuration, regardless of its total velocity (V_T), as the *structural* photon. If two equal velocity 50/50 $_T\alpha$ bisect perpendicularly, that form (or structure), whether it's $V_T = 10^{26}$ m/sec or 1 m/sec, is a structural photon. Those structural photons which fall within the observer's range of perception are *perceptual* photons. Think of the tiny structural union of sperm and egg manifesting as embryo, fetus and finally the visible infant. This perceived form then manifests as child, adolescent, and adult. At death, the person assumes a form we can no longer perceive, even though the structural integrity of the individual remains intact. Like all things, it is structural but not perceived before (perceptual) birth and after (perceptual) death. Although we may choose to argue only the perceptual form is real (here), that doesn't mean the other form(s) doesn't or can't exist somewhere else.

So the perceptual forms have corresponding structural components at total velocities beyond that of the perceived photon. We propose this dual designation to eliminate some of the confusion surrounding the photon which has been variously defined as

- Energy.
- Light.
- Particle-like.
- Wave-like.
- Electro-magnetic.

Casual observers reviewing this plethora of seemingly-conflicting definitions can't help but wonder, "Well, what is it?"

The answer is, the photon is all of these, and by defining it as a *configuration* rather than a "thing", we can explain all its different forms. As long as we keep in mind that the photon is the dual 50/50 combination and that *all* energy forms are perpendicular bisecting pairs relative to there, we'll have no trouble understanding light and the energy forms.

From our discussion of layers and interphases, can you guess where the perceptual (light) photons begin? (Review the table at the end of chapter four.) Because we know we can't see the layers, they (the layers) must function as energy relative to human perception. We also recognize that what we perceive as real is a function of the interphases. Therefore, the source of the perceptual photon must be interphase 8-9 (a 50/50 realm). All other inherent perpendicular layer combinations are perceived as energy, either linear and/or spin.

Now let's talk a bit about the philosophy of light. For the majority of people, light-seeing is the most depended-upon sense, and therefore the one posing the most problems. Science, and religion to some extent, are filled with far more warnings concerning the dire consequences of light loss (darkness) than any other form of energy. A blind person is far more pitied than one who is deaf. The darkness of space is considered the physical manifestation of spiritual deprivation (hell). Fear of darkness is perhaps the greatest fear of all. Think of all the locks, alarm systems, even high illumination quite literally to keep the darkness out. The worst manifestations of physical humankind in the "here and now" (robberies, muggings, murders) and in the multidimensional realm (devils, demons, "things that go bump in the night") are associated with darkness. Insecure large populations create massive cities that function 24 hours a day in an attempt to turn night into day for those who share a fear of darkness. On the other hand, all that is good (enlightened) is considered the epitome. Light is used to describe the best in the physical (body), intellectual (mind), and emotional (spirit) worlds. Remember how quickly the word "light" and its other forms crept into our advertising—light-weight, lite beer? What about describing someone as light-hearted? Christ said, "I (you) are the light of the world." We describe light music, light conversation, light entertainment implying such forms, while "good", aren't lasting because they are *too* easy, *too* enjoyable. The positive light references in our language and in all cultures are as abundant as the human race's religions; it's the energy form which most closely unites science and philosophy. Because of this, let's end this chapter with a look at the transmission of light.

If we consider our sun as our primary source of light, we must immediately recognize that light can't travel as light "packets", otherwise we'd see them. We can't accept a *black* outer space between us and any star or sun and maintain that what we know as *light* travels *in that form* through that void. If there's no light in our room and we say, "There's no light here," then we can't look at similar darkness in outer space and say, "There *is* light there."

Try to dispel any ideas of the visual (perceptual) photon, a pinpoint or spark of light, as indicative of the *only* form of light. In outer space and down to the 8–9 interphase, light *as we know it* doesn't exist. Light as we know it is a mass manifestation. When insufficient $_7\alpha$ are present to register "light", "mass", "rainbow", we don't see them. They may be there, but they aren't here.

What are the ramifications of viewing light and other forms of energy this way; i.e., rotationally? Throughout the book we used terms such as incoherent, coherent, and inherent to describe combinations of $_7\alpha$. Incoherent and coherent are already routinely used to describe different forms of light and its transmission. The light that comes from bulbs is said to be incoherent because it's random, not in phase with itself. Because of this, we can bathe an object in incoherent light and produce two-dimensional photographs of that object. Lasers produce coherent light—light that is "pumped" so it's synchronized in phase with itself:

incoherent light   coherent (laser) light

Using two coherent light sources (lasers) to illuminate an object, we can produce a hologram—a three-dimensional photograph; as you walk around the image you see the sides and back of the object. When you break the glass "negative" of a hologram, each fragment contains the entire image just as each one of our 10^{13} pairs of $_7\alpha$ comprising the perceptual photon has all of its characteristics.

If it's possible to trick our eyes into seeing a three-dimensional object using coherent light energy, we suggest it's possible to trick our other sensory organs to perceive sound, taste, scent and touch via the same mechanism. Rather than using coherent perceptual forms in the light range, one simply uses those coherent forms corresponding to our normal perceptual range of these other senses. The result? A total illusion we can see, hear, touch, taste and smell.

Now if such a total illusion merely demands we expand what we know about light just a little bit more, is there any reason we can't go one step further? If a total illusion can be produced using coherent energy, what do you think *inherent* energy creates? If all is one, where does illusion end and real begin? Is an illusion real because it's here? But if all our senses say "It's real," how can it be an illusion?

What is real? There are countless answers to that question but the best is being in love, being at one (inherent) with the self. Once this occurs, we gain self-confidence and can recognize (perceive) our freedom and potential to expand in all directions. In doing so we seek out that which is the most opposite from ourselves, that which offers us the most expanded view of reality. Our common bond is that we want to be together, yet each one of us maintains our own unique rotation, orientation and identity. We are inherently bound by choice. The energy is love. The result—a mass reality manifesting the purest form of that love.

There are many forms of light and many definitions of those forms, but if we always go back to the basic inherent pair, we can't get lost.

10 | Where Do I Belong?

In the past nine chapters we've spoken of the singular $_7\alpha$ and its combination forms—incoherent, coherent, composite, and inherent—and touched on the multidimensional (multiplanar) realm. Now we're going to expand the concept of the multidimensional $_7\alpha$ to include the human species.

Because all is rotational in nature, we can represent *anything*, not just the $_7\alpha$, in our now-familiar multidimensional form

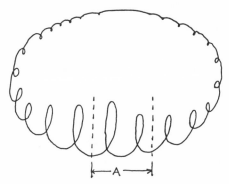

where A represents the limits of any entity in any given reality. Of course, A can vary not only in its length at any place around the spiral but in its position as well.

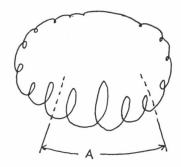

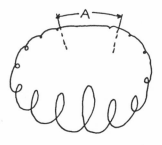

Because the $_7\alpha$ spiral is a model for all that exists within this reality, as it must be if this is to be a true Unified Fields Theory, we can use it to shed some light on the ancient question, "Where do I belong?" Inherent in that question are two others: (1) "Where did I come from?" and (2) "Where am I going?"

Suppose you are a $_7\alpha$. Here you are whipping along at high speed and you encounter the "edge" of a new reality. (Remember, this isn't a case of bombing into this reality because all is continuous.) However, because this new environment looks interesting to you, you decide to enter it. You slow down, maybe change direction. How many times have you driven by a road that intrigues you and then one day impulsively take it to see where it goes? You could have studied a map in advance; you could have asked others; you could have driven up the road further and further each day over a period of days. But you plunge in. There's a dual feeling—one of the unknown ("Will this road dead-end in some terrible place?") as well as one of excitement. ("I haven't been this way before.")

The $_7\alpha$ functions much the same way except that it's the *road* as well as the vehicle on it. On the one hand it knows where it's been and where it's going; on the other, the trip is a brand new experience in this time and place.

We've said before in one way or another that every entity is omniscient (all-knowing), omnipotent (all-powerful), and omnipresent (everywhere at once). If all this "omni" stuff is real, why would any entity bother *doing* anything if it already knows the outcome of its actions and all the circumstances surrounding those actions? The answer is that all entities actually use kind of a limited "omni-all" most of the time. The reason for this is to permit personal discovery and the learning of lessons unique to that entity in that reality. If all chose to instantly know all, existence would be most dull indeed. Did you ever play poker? Suppose you knew exactly what every other player held in his or her hand? It might make you lots of money (It might also get you shot.), but it wouldn't be much fun.

So when the $_7\alpha$ (you) enters a reality it almost always makes a deliberate choice not to *consciously* know whether it will stay here one-trillionth of a second or a billion years, or what it will do in that time. It can appear as a minute track in a high-speed cyclotron or become one with the Sphinx and hang around for eons.

Think of a train capable of infinite speed in an infinite number of directions on an infinite track, and this reality as a valley between other realities, which themselves are between others. When the train enters this reality it may be traveling so fast we never see it. Or it may pass through at a very predictable rate such as the birth, growth and decay of the annual marigold or long-lived giant sequoia, as an amoeba or wildebeest. Or it may move so slowly that relative to *our* life span/motion, it appears not to move at all, like the Washington Monument or the Rocky Mountains. It's still the same train, it's still on its infinite track; it still retains the ability to make any one of an infinite number of choices regarding its speed (the amount of time) and direction (its rate of change) here. The shorter the time, the greater the rate of change relative to this reality, up to the point of passing through so rapidly we're not aware of it at all. The longer it's here, the less change we see. Familiarity breeds contempt; it's harder to see changes in things that are around us all the time.

How far back can you remember? Try this exercise—think of the first event you can remember in this current life of yours. Now back it up one "memory". Where does this leave you? As an infant in your mother's arms? Being born? In the womb? Before that? What if each one of us is more than the biological happenstance of a man and a woman who chose to make love? What if, via inherent omniscience, long "before" our physical births or even conception, we select(ed) our parents just as they select(ed) to "have" us? We all *intuitively* know where we came from but the memory, the *conscious* memory, is difficult to come by.

Why is this, if it's so normal and natural? There's an ancient legend about the depression in the center of the upper lip of most people that says an angel places his/her/its finger there at the time of birth to silence us about our pasts. There's truth in that story: it's difficult to give mass form (memory) to an energy form (spirituality). Relative to what you are now (your "real" or mass form), all other realities are energy-like. Think of trying to remember an early evening dream the next morning. Although some dreams do appear more real than others, it's much, much more difficult to remember and *maintain* that memory than one of a trip to Aunt Harriet's last weekend. Remembering an alternate reality, an energy state relative to mass-dominated here, amounts to trying to hold a moonbeam in your hand.

A $_7\alpha$ is here because it chooses to be here. We're here for the same reason; life isn't a crap-shoot. We're all here because we want the experience of being here. It can't be any other way. No god or supernatural being can place us here, unless we believe we have no free will at all. If we believe we have no free will, what is our reason for being here? To serve some god who's *already* omniscient, omnipotent and omnipresent? How is that different from slavery? Can you honestly imagine the most benevolent, loving god requiring his/her/its subjects give up their free will? That's the worst form of tyranny—it's multidimensional dictatorship. No god worth his/her/its thunderbolts would ever engage in it.

Free will is a very difficult concept to deal with because it means we must accept responsibility for and control over *everything* that goes on in our lives. If we lack self-confidence, that's not an easy thing to do. When things go right it's easier to say we're blessed (by a parent, boss, god, or fate); when things go wrong we often prefer to see ourselves as victims of those same individuals or conditions.

Let's do a little thought experiment in free will. Look at the clothes you're wearing. Did anyone *make* you wear them? Did anyone (thing, being, god) make you eat what you had for breakfast? What about your job? Who you live with? How you view yourself and the world? How you think? What your multidimensional nature is? Is there any aspect of your life you *truly* believe you have absolutely no control over whatsoever? Our question is—If you draw a line somewhere between what you control and what others (including a god) control, why do you draw the line where you do? Would others draw it in the same place at the same time? Will your line always stay in the same place? Think of how some young children are totally terrified of water the first time they see a swimming pool or lake. At that point, they truly believe that water is capable of reaching up and grabbing them, dragging them to some horrible fate. With the confidence that comes from supportive parents and friends and swimming lessons, the *fear* subsides. Now they no longer feel the water is beyond their control; the water demon has become a welcomed and enjoyable expansion of their reality. What happens when the ninety-eight pound weakling who previously saw himself as fair game for every mugger in New York earns a black belt in karate? Does his line between what he can control and what he can't stay in the same place? We have a friend who likes to say, "Nobody knows the answer to that," when he doesn't know or understand something. We can't figure out who or what defines what "that" is. Some of the things he says nobody knows are quite common knowledge outside his rather limited range of experience. Still, he functions as though there is a definite line between what the human race can and cannot know.

Remember asking your mother "Where did I come from?" We can ask the same question regarding the origin of the $_7\alpha$: Where does it come from? The answer is infinite. In most cases the $_7\alpha$ either comes from another reality (which may be much like ours or vastly different) or from the multidimensional realm itself. What's the multidimensional realm like? Imagine yourself bodiless deep in outer space where there's no time and no mass. You have your mind but you have something else too, don't you? Depending on whom you talk to, that "something else" is your energy essence, your spirit, your soul. Because there's no time you're eternal, just like the $_7\alpha$.

Earlier we described how a $_7\alpha$ enters and leaves this reality. If the smallest entity operates in this fashion, then we must be able to also. If we extend our concept of free will to its limit, we must logically recognize death, the departure from this reality, as a matter of choice. This may be the most difficult concept of rotational physics to work through, but it's also the most freeing concept of all. Death isn't the end of all but merely a voluntary exiting of one reality. Your choices remain infinite upon leaving this reality: (1) You may enter another reality, (2) You may exist in the multidimensional realm, (3) You may return to this reality. The latter is called reincarnation.

How do we know this? Up to now we've more or less represented the uniplanar manifestations of $_7\alpha$ as linear "zingers" and spinning "bombs" whipping through space and banging into each other. Although this orientation quite nicely describes how the $_7\alpha$ manifest here (that is, in the layers comprising our perceptual reality), it says nothing about the rotational or continuous nature of the $_7\alpha$ which confers its multidimensionality. Without this, the $_7\alpha$ is no different from any other smallest entity. We've all been exposed to zinging rays and particle bombs in the popular press. Rather than being impressed by such moieties, most people are frustrated. It's most disquieting to view the universe as nothing more than infinite high energy particles raining down upon and through us and colliding all around to create our reality.

Such theory implies we have no part in our own creation. If all is created by countless big and little bangs, piling up mass here and destroying it elsewhere (like the sun) how can we believe our destiny is any greater? Given the current view of physics, is it any wonder nuclear destruction is so terrifying to so many? Nuclear war is merely a technological version of what many particle and astrophysicists say occurs *naturally* all the time. Nuclear war isn't terrifying because it's so alien; it's terrifying because it so closely matches the mass beliefs of what is naturally *real*.

Is it also any wonder, then, when currently surrounded by a workable technological model capable of re-creating what the majority perceive as a natural event, so many now seek a god? Within current *available* physics no one is saying there's more beyond our here, what we see, touch, hear, smell, and feel. Particle physicists say things "disappear" and "appear" implying a black meaningless void, a hell of nothingness, sucking black "holes", blinding quarks, monotonous pulsars. Religion more clearly aligns with our intuitive beliefs that there's a "something" beyond here which is invariably described as much better (heaven) or much worse (hell). Many religions support a belief that some massive change occurs "there". Some preach that those who are "good" (a highly relative concept at best) are rewarded with bliss and eternal peace, whereas those who are "bad" will suffer horrible consequences. However, suppose, like the $_7\alpha$, the only change that takes place upon death (i.e., the exiting of a reality) is the loss of the physical body—a change in *form*. If you have a problem you haven't resolved here, it will still be with you there. Think of all the cultures that have taboos against suicide; the act of taking one's own life is invariably seen as an attempt to run away from something—a poor relationship, weak self-image, problems, pain. It isn't the death that results in the sin (separation); it's the individual's blatant statement that he or she doesn't realize it's impossible to run away from one's self. Escape suicide is like moving into a different room to get away from your own body odor. All that happens is that you have to deal with the same problem in different surroundings.

Why wouldn't we be the same (except for form) here as there? If we really believe we're truly unique, what logic can support that, upon death, we become something totally different or alien? Even if you picture yourself in a heaven or hell complete with wings and halo, or horns and pointed tail, is there ever any doubt in your mind, you're you? Or think of all the dreams you have where you know and do things you don't know and do here, or those where you appear younger, older, or even the opposite sex. Yet in your mind, regardless of the form you take, there's never any question who you are.

One of the purposes of rotational physics is to show not only that there is a there but how that there functions relative to here; that regardless of what happens here, a there full of infinite other potential heres does and always will exist. We may slightly alter a credo from the Christian religion (certainly implying no disrespect): "Glory be to the father *(my multidimensional self there)*, to the son *(my earthly self and potential self in this reality)*, and to the holy spirit *(that energy essence which unites the two selves with a composite bond). (My)* World *(reality)* without end. Amen. *(So be it.)*"

We may view this same credo in terms of physics where we exalt in our ability to manifest our full potential multidimensionality in this reality and all forms between. Doesn't Christianity view God the Father as the epitome of all that's there, Christ as the very best here, and the Holy Spirit as the bridge between the two—and the rest of mankind? Or take it one step further viewing the father as energy, the son as mass and the spirit as the intermediate energy-mass state. Now that most elegant credo enjoins us to rejoice in *all* the different forms inherent in each individual's potential. There are many other trinities both in nature and within the arts and sciences of all cultures, all intuitively acknowledging humankind's continuity with all that is.

Let's see how fears regarding multidimensionality, the presence or lack of something beyond here, can affect an entire world. The current status of nuclear weaponry and its politics serves as a fine example. If one is convinced a there exists, one feels no driving need to get there to prove it. We only feel the need to prove those things we're unsure about. Obviously those who believe they can solve their political (belief) differences by destroying all or part of their reality, function under the same poor logic as those who hope to escape themselves via suicide. If we blast ourselves to smithereens here, we'll simply have to deal with the same problems in a different form there. So if we believe in a there in a non-religious context, nuclear war or armament makes no sense. We can solve any problems right here, using talk, rocks, sticks, guns, or whatever mode of communication politically in vogue.

If we believe in there as the domain of a god, there is also no reason for nuclear war and armament. Remember David and Goliath? David didn't find it necessary to build a nuclear slingshot or involve thousands. He trusted in his belief; because he didn't *have* to prove anything to himself, he freed himself to use all the multidimensional (god-) power at his command. We can't help wonder about those leaders who simultaneously profess deep religious convictions and arm for war. It would seem that if one truly believes in an omnipotent, omniscient, omnipresent deity, such armament would be blasphemous.

Obviously, then, if we believe there is a there, whether via religious, scientific or philosophical convictions, nuclear war is meaningless. There's no reason for it. We can rationalize, pointing to man's competitive nature, that some might enjoy the intellectual achievement of constructing a bomb capable of destroying the earth. We can even go so far as to say that, once one individual has that capability, anyone who fears him or her or lacks confidence in their ability to resolve differences some other way, will "need" a bomb too. However, once all those who believe they need a bomb that can destroy the world have such a bomb, what's the reason for each then needing two, three, a hundred or fifty thousand such bombs?

Fear. Such people are terrified there is no there. They're terrified that here is all there is. They don't believe their gods and they don't believe their leaders whose actions are dominated by fear-based defense responses. Let's examine two events that are currently occurring in the Soviet Union: one is a widespread acceptance within the scientific community, military and general population of many events the majority of Westerners pooh-pooh as "psychic bunk". The other event is the Soviet concentration of their defense program on more traditional forms of weaponry. Although we can say the Soviets have taken a very hard line against a religious god-based there, their interest in and acceptance of psychic phenomena tells us they (a) believe in a there, and (b) are not afraid of it. In many ways their knowledge and acceptance of a there permits them to concentrate more on being here. Thus those who feel differences must be resolved via force are quite comfortable with a defense program many Westerners consider technologically primitive. However, such weaponry is quite adequate for those who wish to resolve their differences *and* remain here.

Let's look at the situation in the U.S. We're desperately searching for evidence of a there. At first we thought science and technology would give it to us, bring it back from space, eliminate disease, give safe, clean energy. But all those efforts seem to bring are more and more expensive equipment, more incomprehensible terminology, and a greater sense of alienation. Many then abandoned science for religion, all sorts of religions. We worship anything and everything—gods, devils, health, youth, power—hoping to find that here isn't all there is. Then there are those who think nuclear war is the answer. These people so desperately want to know, to believe there is there, they'll blast themselves there if religion can't provide the answer. And because they're so frightened of the unknown, like frightened youngsters they want to take all of us with them—just in case.

If we remove the fear of the unknown, the fear that here is all there is, then that minority which chooses to prove the existence of there by blasting an entire reality into another loses the reluctant but highly frustrated audience that *helps* keep their erroneous belief alive. The best way to cause something to cease to exist is to *ignore* it. It's a well-known principle of behavior that rewarding *or* punishing behavior keeps it going longer than if we do nothing at all. Why is this? This occurs because the vast majority of the time those displaying the behavior really want *attention*; they don't care whether it's our support or condemnation; either response stimulates them to continue the behavior. Think of a child having a temper tantrum—what's the best and fastest way to stop not only that one but decrease the likelihood of the child using that form of behavior again? Ignore it. Remember the old peace-nik slogan: What if they held a war and nobody came? It's the same thing. Those who are rabidly anti-nuke or anti-war keep the process going as much as those who are equally pro-nuke and pro-war.

Relative to answering the age-old question of continuity, immortality, and multidimensionality, nuclear warfare is current technological society's answer to the religious fervor of the past. Its proponents feel they have shunned religion (superstition) and evolved to this point. They truly believe science has given them nothing to indicate a heaven or a there exists. They're so terrified of death they would propel themselves into it prematurely to eliminate the pressure; so terrified, they feel they must take others, *many* others, with them. They are at a point where their only question is, "What else can we do?"

Fear is the worst, most detrimental weapon we use against ourselves and others. If there's no fear, there's no need to fight. Most of current science focuses on here while many religions talk about there; each portrays the other as separate and distinct, serving only to reinforce our belief that all is *not* continuous.

How about some proof? If all is continuous and therefore omniscient, omnipotent and omnipresent (even if we choose to limit ourselves in this reality), there must be some readily available evidence of this other "side" of our nature. Perhaps one thing that makes here here is our sense of time. In our highly technological society we now have throw-away watches of incredible accuracy: but how real is time?

Let's talk a bit about time and its simultaneous (rotational) nature. Thoreau once said, "Time is but a stream I go a-fishing in." Although we marvel at the ability of our electronic quartz crystal watches to keep *exact* time, scientists have been telling us for 100 years or so that time is highly relative. Remember Einstein's famous theory about the identical twins where one brother journeys in a space ship at the speed of light? If he travels for ten years and returns to earth, he notices his earthbound brother has more wrinkles, a slightly bulging mid-section, and some grey hairs. His twin has aged ten years, but our traveler hasn't aged *one minute*! Or, consider the limerick about an amazing English woman:

> *There was a young lady from Wight,*
> *Who could travel much faster than light.*
> *She went away one day,*
> *And in an Einsteinian way,*
> *Returned the previous night.*

Simultaneous or rotational time might be our biggest single proof of multidimensionality for the $_7\alpha$, for larger rotations that occur within atoms and galaxies, and ourselves. Let's look at a quick proof of the concept.

The next time you speak to someone, pay attention to the process. When you start talking, do you have a specific idea how your first sentence is going to end; i.e., its *exact* wording? Unless you memorize and repeat it *ver batim*, you don't. Now, *how* did that sentence come into being? The agonizingly slow response of neurons within the body tell us the brain can't put it all together in the short time it takes to formulate a thought and verbalize it. It has to come from somewhere else and that somewhere is what we call the future. That is, we're able to complete our sentences because we know what we're going to say, even if we're not consciously aware of it. Our conversations with others aren't a process of our first thinking "I'm going to say to Helen, 'My, that's a lovely dress you're wearing'," followed by the verbalized, "My, that's a lovely dress you're wearing." Generally if we do feel the need to consciously create conversation *before* we verbalize it, it's because something about the situation or the other person makes us question our ability to easily pull thoughts/words from "there" when we need them. Not surprisingly, this lack of self-confidence is merely another form of fear.

When parapsychologists help people see past life experiences, why can most people "relive" portions of those lives in vivid detail? For most of us, our memories of things past, even what happened yesterday, is limited to say the least. However, suppose that past life is occurring simultaneously in a reality parallelling this one. Wouldn't that account for what occurs better than "regressing" to a time that preceded our present one? It's the difference between moving from one room into another versus trying to remember what went on in that same room years ago.

In this chapter we've taken some of what we know about the $_7\alpha$ and applied it to human nature. We did this simply because what holds for the least must hold for the greatest. If our theory holds for the $_7\alpha$, it *must* hold for everything—atoms, molecules, rocks, plants, animals, humans, planets, solar systems, galaxies, universes, mega-universes—both here and there.

The $_7\alpha$ comes into this reality at high speed but relatively uncharged ($\pm$). It gains relative charge ($+$ or $-$) here as it interacts with other entities. Once it achieves its relative charge here, it may choose to modify its form as a function of others of its kind. If it changes form, it can then stay in that state or alter it as it sees fit. It may stay here for what we may call an infinite length of time or it may leave immediately. So it is with us. A child may die at birth, leaving its parents here, but is that child really "lost?" An old person dies, leaving all his or her memories and works *behind*; does that mean life is over? Where do *you* draw the lines between living and being? Aren't we more than physiological entities bounded by form and time?

This has been a most important chapter in our theory of rotational physics because it shows how what applies to the $_7\alpha$ applies to each and everyone of us. Not only are we uni- and multidimensional beings of infinite potential, we are capable of forming an infinite number of relationships. When our two $_7\alpha$ of different total velocities form an unstable pair that eventually separates, do you feel either $_7\alpha$ is "wrong?" When two $_7\alpha$ of equal velocity intersect perpendicularly at their opposite ends (a single negative intrinsic charge of one aligns with a single positive intrinsic charge of another) and form a weak bond which breaks and gives rise to a most stable composite, do you feel they should have tried harder to stay perpendicularly aligned? Think of a sound stable marriage as an inherent pair and friendships as stable mono- or bipolar combinations. In order for the combinations to remain intact, each participant must view the other as equal; not the same, just equal in total potential. That has to be the starting point. As long as our component $_7\alpha$ have equal total velocity, it makes no difference in the stability of the bond what *form* that velocity takes. One component may be all spin and the other all linear motion, or any combination of the two; all forms carry equal weight. If we think Americans are better, whites are better, men are stronger, women more intuitive, or anything that denotes a relationship based on inequality rather than equality with a difference, nothing lasting can come from that relationship. It's a union built on the weakest, the most sandy of all foundations.

We're sure you can think of many more examples of how rotational physics applies to your own experience. Take the time to understand the basics, then the answer to the question "Where do I belong?" will be quite clear. Where do you belong? Exactly where you are. Otherwise you wouldn't be here.

Epilogue

This small and simple book is a beginning. It's also an end—
an end to an era that has brought some solutions to the human race
but also many problems. There are presently 50,000 nuclear weapons on the planet. Has anyone honestly every figured out why?
For security purposes? No. Security is a feeling each individual
does or doesn't create within him or herself. Weapons don't create
security; they create insecurity. And nuclear weapons create the
worst form of insecurity: group cowardice and fear.

We can't use any form of weaponry to conquer fear because
the weapons themselves are indicative of our fear. We can't use
fear to fight fear. It's like fighting fire with fire—all we get is more
fire and more destruction. Those who rationalize this is the *only*
way are afraid of the alternative. The alternative (which they may
define as communism, capitalism, socialism, or some other -ism)
isn't a political structure or group of people at all. The alternative
we all fear at one time or another is the unknown. If the only thing
we recognize about the unknown is that it's wrong—the wrong skin
color, wrong religious beliefs, wrong language, wrong political
structure—rather than different, our fear invariably leads to defen-
sive reactions.

If we want to conquer our fear of our fellow humans, we must
want to know and understand them as equals. We must do this
ourselves. We can no longer cop out by saying we're not responsible
for our ignorance and the resultant fear and weaponry. We all did
this to ourselves, so it's up to us to undo it. How can we undo it?
Remember one of the first things we said about the $_7\alpha$ was that it

takes at least two $_7\alpha$ for there to be any meaningful activity. If there's only one, it can go through dizzying changes but it makes no difference. The same is true about ideas and behavior. If a negative idea or behavior is simply ignored, it goes away. The more vehemently one opposes it, the more one reinforces its existence.

Some may propose that arguing against something we perceive as bad or wrong is a matter of principle. That's true; but it does nothing to create any change. Far better to ignore what is now and focus on what we want. Again, think of our $_7\alpha$: When two combine having different total velocities, it doesn't matter which one is stronger. The result is always highly unstable. All it takes is commitment, a personal commitment to *create* a better world, not *destroy* what we believe is wrong in this one. The era of "Let George (or Harry or Ilene or some president) do it" is over. We must each *do it* or it won't be done.

In this *Primer* we've given you a glimpse of what's to come. This isn't an academic exercise describing meanderings of some hypothetical particle we can view via arcane instrumentation. The $_7\alpha$ is a part of us all because it's infinite in its capabilities, both infinitely small and large at the same time. Scientists and philosophers alike have searched for centuries for some bond that makes us all part of one another, that makes us belong. Most religions espouse we're all part of all that is but provide no proof other than presenting their beliefs as a matter of something called "faith". If we can look to science and see something reflected in philosophy and vice versa, it becomes its own proof. If we can look at theory and recognize, "Hey, that's what happens to me when I _____," or "I've had that up tight (high spin) and strung out (high linear energy) feeling at work!", the idea of a unifying entity doesn't seem so academic any more.

In this final chapter of our first step through our unifying theory, let's look at the eloquent and beautiful I Corinthians 13 and see how our rotational theory fits in:

"If I speak in the tongues of men and of angels, but have not love, I am a noisy gong or a clanging cymbal. *[If I have knowledge of all that is, but am not at peace with what I am, I am incoherent.]* And if I have prophetic powers, and understand all mysteries and all knowledge, and if I have faith so as to remove mountains, but have not love, I am nothing. *[If I recognize my omniscience, omnipotence and omnipresence there, but not my free will, my choice to be here, I am nothing.]* If I give away all I have, and if I deliver my body to be burned, but have not love, I gain nothing. *[If I give up all I have and am here to prove a there exists, all I prove is my ignorance of the simultaneous nature of here and there.]* Love is patient and kind; love is not jealous or boastful; it is not arrogant or rude. Love does not insist on its own way; it is not irritable or resentful; it does not rejoice at wrong, but rejoices in the right. Love bears all things, believes all things, hopes all things, endures all things. *[Energy, the multidimensional me, is.]*

Love never ends; as for prophecy, it will pass away; as for tongues, they will cease; as for knowledge, it will pass away. For our knowledge is imperfect and our prophecy is imperfect, but when the perfect comes, the imperfect will pass away. *[Energy is multidimensional. The uniplanar teachings and mass of this reality are bound by time and space. When we are no longer here ("imperfect") in mass form, we are there as energy.]*

When I was child, I spoke like a child, I reasoned like a child; when I became a man, I gave up childish things. *[When I was young I was more aware of my multidimensional nature, but I lost that awareness as I became more involved in a mass reality based on time and space.]* For now we see in a mirror dimly, but then face to face. *[As fully integrated parts of here, our awareness of there is primarily intuitive, whereas as children there is equally real, albeit different, to us.]* Now I know in part, even as I have been fully understood. *[Here I am somewhat aware of my omniscient self, just as my omniscient multidimensional self is aware of me.]* So faith, hope, love abide, these three; but the greatest of these is love." *[So awareness of my multidimensional nature as neither good nor bad, my choice to be here, and the always-present energy to change is the basic foundation of all that is. But the greatest of these is the energy, the power to change.]*

Like all things, this beautiful passage from scripture has infinite meanings and therein lies its beauty, for it can speak to each one of us in our own language. The nature of anything that brings unity isn't that it's so highly defined and specific it encompasses everything: it's the glorious versatility and adaptability of an entity that enables it to unify all.

ALL.

Acknowledgements

The Primer of Rotational Physics could not have been written without some very unique and highly specialized help. Much of the material in this book and the ones to follow is extracted from thousands of pages of scientific and philosophical material amassed since April 30, 1982, via the parapsychological method called automatic writing. Our major contributor to the work is Albert Einstein, although Lewis Carroll (aka the mathematician Charles L. Dodgson), René Descartes and Jesus Christ also participated.

"Hold on," you say. "You mean to tell us that this UFT you so loftily talked about is produced through some kind of hocus-pocus?" No, there's no magic about it. Many people have done such things and anyone can do it. It's not magic, occult, or even miraculous. Furthermore because everything is connected, why wouldn't we all have access to all even if our forms are different?

However, if you have difficulty accepting a paranormal origin for the text, you may prefer to believe we made it up. That's okay; we're not here to force our beliefs on you. The question is: Does it make sense? If it does and you believe in the paranormal, you're beliefs are now expanded into the field of physics. If rotational physics makes sense and you don't believe in its origin, you've gained a new understanding of physics and perhaps the slightest inner inkling you could create (or make up) something like this, too. Either way, you've grown and expanded and become more unified in the process.

If you've read the book and it doesn't make sense, well, maybe it's just not time; maybe we're not ready for each other. Just as weeds are often described as out-of-place flowers and strangers as friends we've yet to meet, so our exposure to new ideas like that to paintings, music and other life experiences initially having little relevance simply means they have little or no relevance for us *now*. This in no way means these experiences are wrong or inferior, any more than we are wrong or inferior because we can't appreciate them. The nice thing about rotational physics, regardless where it comes from, is that it lets us know we can never lose anything unless we want to; whenever we're ready, whatever we need is always there.

Our second acknowledgement is to you, our readers. By virtue of your participation you have expanded the concepts in your own unique way. As Emily Dickinson so aptly noted, words are no more dead entities than anything else in creation. They don't pass out of our existance the moment they are said or read. Indeed, it is just the opposite; it is our reading and speaking that gives them eternal life. For this, we thank you.

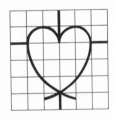